Proprioception

Edited by José A. Vega and Juan Cobo

Published in London, United Kingdom

IntechOpen

Supporting open minds since 2005

Proprioception
http://dx.doi.org/10.5772/intechopen.92928
Edited by José A. Vega and Juan Cobo

Contributors
Alejandra Vasquez-Rosati, Carmen Cordero, Kaviraja Kamatchi, Wangdo Kim, Juhani Partanen, Urho Sompa, Miguel Angel Muñoz-Ruiz, Alexander Vladimirovich Zakharov, Elena Viktorovna Khivintseva, Alexander Vladimirovich Kolsanov, Alexander Vladimirovich Yashkov, Vasiliy Fedorovich Pyatin, Giridharan Vaishnavi, José A. Vega, Juan Cobo, Juan L. Cobo, Pinar Gelener, Ramadan Özmanevra, Gözde Iyigün, Sonsoles Junquera, José Martín-Cruces, Antonio Solé-Magdalena, Olivia García-Suárez, Teresa Cobo

First published in London, United Kingdom, 2021 by IntechOpen
IntechOpen is the global imprint of INTECHOPEN LIMITED, registered in England and Wales, registration number: 11086078, 5 Princes Gate Court, London, SW7 2QJ, United Kingdom
Printed in Croatia

British Library Cataloguing-in-Publication Data
A catalogue record for this book is available from the British Library

Additional hard and PDF copies can be obtained from orders@intechopen.com

Proprioception
Edited by José A. Vega and Juan Cobo
p. cm.
Print ISBN 978-1-83968-069-4
Online ISBN 978-1-83968-070-0
eBook (PDF) ISBN 978-1-83968-074-8

For EU product safety concerns:
IN TECH d.o.o., Prolaz Marije Krucifikse Kozulić 3, 51000 Rijeka, Croatia, info@intechopen.com or visit our website at intechopen.com.

Meet the editors

José A. Vega obtained a Ph.D. in Medicine from the University of Oviedo, Spain. He completed his postdoctoral training at the Universities of Brno and Prague, "La Sapienza" and "Tor Vergara" of Rome, and Harvard University. Currently, Dr. Vega is Professor of Anatomy and Human Embryology, and head of the Sensory Organs and Peripheral Nervous System (SINPOS) research group at the University of Oviedo. He is also an Associated Researcher at the Autonomous University of Chile. He has taught at the Universities of Messina, "Federico II" of Naples, Rome "La Sapienza" and Rome "Tor Vergata," Sassari, Barí, Buenos Aires, and CEU-San Pablo in Madrid. Dr. Vegas has co-authored more than 350 articles and 50 chapters in books mainly devoted to the peripheral nervous system.

Juan Cobo graduated in Medicine and Surgery from the University of Zaragoza, Spain, and obtained a Ph.D. from the University of Oviedo, Spain. He completed his postdoctoral training at the University of Chicago, USA. Currently, Dr. Cobo is a Professor of Orthodontics and head of the master's program in Orthodontic and Dentofacial Orthopedics, University of Oviedo. He has co-authored more than 150 articles, four books, and thirty chapters in books mainly related to orthodontics and the peripheral nervous system.

Contents

Preface

Proprioception represents a conspicuous part of the somatosensory system. The actual state of the art in the study of proprioception clearly demonstrates that it is a field of continuous active and innovative research. A large body of evidence has been accumulated in the last years concerning the characteristics of proprioceptors and the peripheral sensory organs connected to them (i.e., muscle spindles and Golgi's tendon organs). Furthermore, the contribution to proprioception of mechanoreceptors situated in the joint capsule and skin is being actively explored.

Proprioception is a collection of reviews and new information on mechanosensitivity. We hope it will be of interest for investigators in basic research, as well as for clinicians with an interest in proprioception.

In the opening chapter, "Structural and Biological Basis for Proprioception," the editors, José A. Vega and Juan Cobo summarize and update the basics of the topic, knowledge on muscle proprioceptors (i.e., muscle spindles), mechanisms and molecular bases for mechanotransduction in muscle spindles, pathways of proprioception, and the clinical importance of proprioception. This chapter provides the common foundations that make it easier to read the subsequent chapters of the book.

The second chapter, "Proprioception and Clinical Correlation" by Pinar Gelener, Gözde İyigün, and Ramadan Özmanevra, focuses on the anatomy, motor control, and postural control related to proprioception with neurologic clinical correlation. It also contains information about the changes in joint proprioception after orthopedic surgeries.

The following two chapters discuss proprioception in the lower extremities. The third chapter, "Recording of Proprioceptive Muscle Reflexes in the Lower Extremity" by Juhani Partanen, Urho Sompa, and Miguel Muñoz-Ruiz, analyzes the proprioceptive reflexes, especially the H-reflex, and recommends recording of this reflex with an EMG needle electrode to perform accurate diagnostics.

The fourth chapter, "The Knee Proprioception as Patient-Dependent Outcome Measures within Surgical and Non-Surgical Interventions" by Wangdo Kim, approaches the problems of proprioception in patients undergoing knee surgical procedures. The authors develop an "evidence-based medicine" design and describe the steps to identify measurable invariants in the knee proprioception system and develop a mathematical framework for outcome measurement within the knee.

The proprioception of the muscles of the head is still an open matter since the nerves driving the information to the central nervous system are not definitively identified and most cephalic muscles lack muscle spindles, which are the true proprioceptors. The fifth chapter, "Proprioceptors in Cephalic Muscles" by Juan L. Cobo, Sonsoles Junquera, José Martín-Cruces, Antonio Solé-Magdalena, Olivia García-Suárez, and Teresa Cobo, explores the presence of alternative or atypical proprioceptors in head muscles. The authors identify three basic types of possible atypical proprioceptors based on the expression of PIEZO2 and ASIC2 mechanoproteins and establish the relative densities of each type in the muscles analyzed.

The sixth chapter, "Proprioception Impairment and Treatment Approaches in Pediatrics" by Kamatchi Kaviraja, introduces the new aspect of proprioception sense and its dysfunction. The author focuses on children with behavioral problems in which early identification and intervention play major roles in improving the ability and development of the proprioceptive senses.

The seventh chapter, "Proprioceptive Perception: An Emergence of the Interaction of Body and Language" by Alejandra Vasquez-Rosati and Carmen Cordero-Homad, provides a systemic perspective of human behavior, which reformulates the concept of effective behavior and cognition that derive from the classical vision of neuroscience and psychology based on the Cartesian reductionist functionalist paradigm. This chapter answers the question of how proprioceptive perception affects human beings' experience of being different from others and from the environment and it explains how this phenomenon modulates.

The chapter "Nomophobia Kids and Proprioception" by Giridharan Vaishnavi analyzes the impact of nomophobia (*the soreness or tension as a result of the non-availability of a cellular telephone, a personal laptop or any some other digital verbal exchange device*) on proprioception, especially in kids. Children using smartphones for a long time experience a great impact on the sensorimotor function with a deficit in proprioception.

In the last chapter, "Proprioception in Immersive Virtual Reality," Alexander Vladimirovich Zakharov, Alexander Vladimirovich Kolsanov, Elena Viktorovna Khivintseva, Vasiliy Fedorovich Pyatin, and Alexander Vladimirovich Yashkov drive the readers to a new, innovative, and almost futuristic projection of proprioception: its connection to virtual reality (VR). Under the conditions of VR, a variety of multimodal sensory experiences can be obtained, and in the opinion of the authors it is necessary and urgent to create immersive explicit environments to bring the full potential of VR technology powers. Activation of the proprioceptive system, coupled with the activation of the visual analyzer system, allows achieving sensations of interaction with VR objects, identical to the sensations of the real physical world. These new devices must be readily available for use in routine medical practice, customizing the rehabilitation process for different pathologies.

Each of the book's chapters represents the efforts of diverse specialists to analyze and review recent findings on several aspects of proprioception. We thank the authors for their collaboration, their effort, and their competence. Finally, we express our thanks to Mr. Josip Knapic and IntechOpen for their invaluable support and editorial assistance. We are also very grateful to the professors who served as reviewers for the chapters. We hope this book will be of interest for investigators in basic research as well as clinicians with an interest in proprioception.

José A. Vega
Departamento de Morfología y Biología Celular,
Grupo SINPOS,
Universidad de Oviedo,
Spain

Facultad de Ciencias de la Salud,
Universidad Autónoma de Chile,
Chile

Juan Cobo
Departamento de Cirugía y Especialidades Médico-Quirúrgicas,
Universidad de Oviedo,
Spain

Instituto Asturiano de Odontología,
Oviedo, Spain

Chapter 1

Structural and Biological Basis for Proprioception

José A. Vega and Juan Cobo

Abstract

The proprioception is the sense of positioning and movement. It is mediate by proprioceptors, a small subset of mechanosensory neurons localized in the dorsal root ganglia that convey information about the stretch and tension of muscles, tendons, and joints. These neurons supply of afferent innervation to specialized sensory organs in muscles (muscle spindles) and tendons (Golgi tendon organs). Thereafter, the information originated in the proprioceptors travels throughout two main nerve pathways reaching the central nervous system at the level of the spinal cord and the cerebellum (unconscious) and the cerebral cortex (conscious) for processing. On the other hand, since the stimuli for proprioceptors are mechanical (stretch, tension) proprioception can be regarded as a modality of mechanosensitivity and the putative mechanotransducers proprioceptors begins to be known now. The mechanogated ion channels acid-sensing ion channel 2 (ASIC2), transient receptor potential vanilloid 4 (TRPV4) and PIEZO2 are among candidates. Impairment or poor proprioception is proper of aging and some neurological diseases. Future research should focus on treating these defects. This chapter intends provide a comprehensive update an overview of the anatomical, structural and molecular basis of proprioception as well as of the main causes of proprioception impairment, including aging, and possible treatments.

Keywords: proprioception, muscle spindles, mechanotransduction, ion channels, proprioceptive pathways, spinocerebellar tracts

1. Introduction

Proprioception is a wider sense, that include position and movement of parts of the body relative to one another, and the force and effort associated with muscle contraction and movement. But properly the term *proprioception* applies for the sensory information contributing to sense of self position, whereas *kinesthesis* refers of sense of movement. The first one is regarded as an automatic function and unconscious in contrast with the second one considered as conscious. In the words of Kröger and Watkins [1] "*Proprioceptive information informs us about the contractile state and movement of muscles, about muscle force, heaviness, stiffness, viscosity and effort and, thus, is required for any coordinated movement, normal gait and for the maintenance of a stable posture*". This information travels to the central nervous system, but differently to other components of somatosensitivity, a great part of the proprioceptive sense does not reach consciousness. This is probably due to suppression as a consequence of the motor signals [2] or inhibitions along somatosensory pathways [3]. The precise knowledge of the pathways of proprioception, especially those of conscious proprioception, are of capital interest to better understand this sense. The techniques of neuroimaging are providing new insights about the cerebral process of proprioception.

Proprioception originates by the activation of proprioceptors at the periphery. Proprioceptors are a subset of mechanosensory neurons that provide afferent innervation to specialized sensory organs located inside the muscles and tendons, but probably also in joint capsules and ligaments, and the skin. According to Proske and Gandevia [4, 5] the sense of "*proprioception is achieved through a summation of peripheral sensory input describing the degree of, and changes in, muscle length and tension, joint angle, and stretch of skin*". In fact, the proper definition of proprioception coined by Sherrington in 1906 ("*In muscular receptivity we see the body itself acting as a stimulus to its own receptors—the proprioceptors*") suggest that the body contains different kinds of proprioceptors. Here we have focused on muscle spindles and Golgi's tendon organs. Especial interest was done on the mechanisms of mechano-transduction and the ion channel in this process.

Proprioception is impaired in some physiological and pathological situations. It will gain interest in the coming years due to the aging of population: the deficit of proprioception is associated with the increased frequency of falls in the elderly [6–8]. Furthermore, several diseases, especially some neurodegenerative disorders, course with proprioception deficits [9, 10] which treatment require a better knowledge of the molecular aspects of proprioception and new active research.

This chapter is aimed not to perform a Review on all the different aspects of proprioception but just to review some general and recent advances in proprioception. We intend to provide the readers of this book with an up-to-date appraisal of the structural and biological basis of proprioception. There are excellent reviews on the topic [4, 5, 11, 12] and we forward the interested to them. Robert W. Banks' extraordinary paper (2015) [13] masterfully sums up the history of knowledge of muscle spindles. Likewise, the recent reviews Kröger [14] Kröger and Watkins [1] are mandatory.

2. Proprioceptors

The peripheral receptors of proprioception are located in tissues around the joints, including skin, muscles, tendons, fascia, joint capsules, and ligaments [15] which contains different morphotypes of mechanoreceptors [16]. It is currently believed that proprioception is not generated by a single receptor, but by multiple of receptors. In any case proprioception has been related with sensory receptors localized in the muscles while kinesthesis has been more associated with joint and cutaneous receptors [17–19]. Nevertheless, the historically regarded as true proprioceptors are muscle spindles and Golgi's tendon organs.

The *joints mechanoreceptors* are Ruffini-like and Pacinian corpuscles which signal joint movement but not movement direction or joint position [15]. Regarding *cutaneous receptors* four kinds of mechanoreceptors are present in glabrous skin (Meissner's corpuscles, Pacinian corpuscles. Ruffini corpuscles and Merkel cell-neurite complexes) [20–22]. A definitive role of cutaneous mechanoreceptors as proprioceptors has not been definitively established [23–25] although it is possible a convergence between cutaneous and muscle afferents at the spinal cord and thalamic levels.

But independently of the modest contribution of cutaneous and articular mechanoreceptors to proprioception, the main stretch-sensitive receptors are muscle spindles found in most, but not all, skeletal muscles. For instance, they are absent form most cephalic muscles [26, 27]. Interestingly, muscle spindles are more abundant in muscles in which the precision of movements must be accurate. On the other hand, the main tension-sensitive receptors are the Golgi's tendon organs, located at the ends of muscle fibers [28, 29]. These two sensory organs respond to changes in mechanical conditions, namely in muscle length (muscle spindles) or in actively generated force (Golgi-tendon organs) but both are contraction receptors.

2.1 Muscle spindles

Vertebrate muscle spindles are complex sensory organs that have both sensory and motor innervation. Each muscle spindle receives at least one sensory fiber that innervate specialized muscle fibers denominated intrafusal fibers. These intrafusal fibers also receive motor innervation by γ-motoneurons [30, 31]. Structurally, they are encapsulated mechanoreceptors, and functionally are slowly adapting-loth threshold mechanoreceptors [5].

Muscle spindles are highly variable in number from none in most cephalic muscles (see [27]) to numerous in lumbrical or deep neck muscles [32, 33]. These differences are attributed functional muscular demands of muscles but the number of muscle spindles per motor unit is rather equal [34]. Also, no topographical differences in muscle spindles between mono- and multiarticular muscles were noted [35].

Within the connective capsule that delimits each muscle spindles there are the intrafusal fibers and the periaxial space filled with a fluid. Three zones can be differentiated at the muscle spindle: the central or equatorial zone, the juxta-equatorial zone, and the terminal or polar zone; small segments of the intrafusal fibers can be found outside the poles of the muscle (**Figure 1**).

The intrafusal muscle fibers. Banks and co-workers [36] established that mammalian muscle spindles regularly contain three types of intrafusal muscle fibers. Based on their morphology and the arrangement of nuclei in the equatorial zone they fall into two main categories: nuclear bag fibers and nuclear chain fibers. Bag

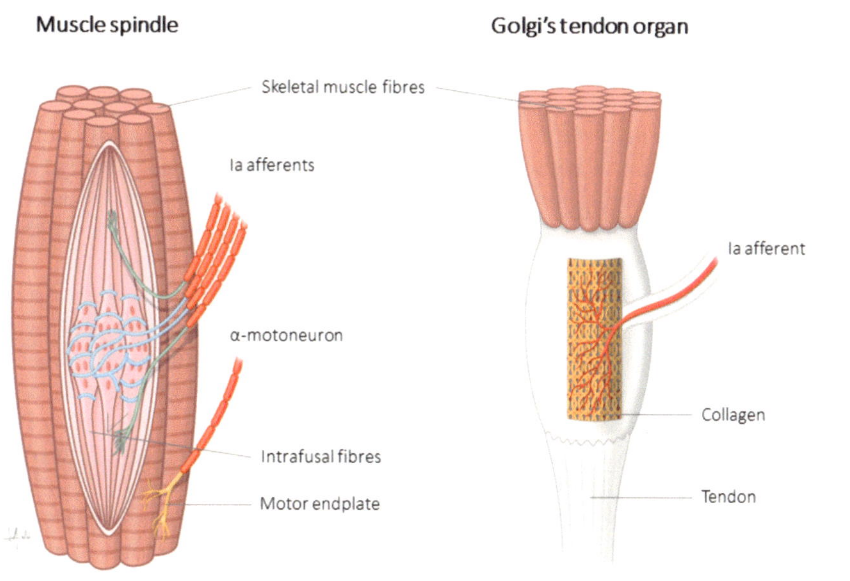

Figure 1.
Schematic representation of a muscle spindle and a Golgi's tendon organ. Muscle spindles are capsulated mechanoreceptors that consist of intrafusal muscle fibers (bag1, bag2 and chain), a periaxial space filled with a fluid, and a connective capsule. They are supplied by Ia (blue) and II (green) afferents. Golgi's tendon organs are capsulated mechanoreceptors that consist of collagen fibers and type Ia afferents (red).

and chain fibers differs in structure, histochemical profile (myosin type, ATPase activity [37–39]) and functional properties. Bag fibers are greater in diameter and length than chain fibers, extend outside of the capsule, and they can be subdivided into bag1 and bag2 types (see for a review [40]). Human contains on average 8–20 intrafusal fibers and can lack bag1 or bag2 fibers [37].

As mentioned previously, muscle spindles are stretch detectors, i.e. "*they sense how much and how fast a muscle is lengthened or shortened*" [41]. Accordingly, when a muscle is stretched this change in length is transmitted to the intrafusal fibers which are in turn stretched. And to respond appropriately intrafusal fibers are double innervated by afferent sensory neurons and efferent motoneurons.

Sensory innervation. "*Just as the number of sensory endings varies from spindle to spindle, even within a single muscle, so also does the number of motor axons supplying those spindles*" [13].

There are two types of afferents that innervate intrafusal fibers: primary (type Ia) and secondary (type II) endings which differ in their axonal conduction velocity [13].

Each muscle spindle receives only one **Ia afferent** surrounding like a dock the equatorial zone of all the intrafusal fibers (spirals or annulospiral endings) (**Figure 1**). When spiral endings deform detect changes in length of the muscle. Primary afferents are sensitive to dynamic stretch, have irregular spontaneous or volitionally maintained discharge, and exhibit an off-response at the point of relaxation (i.e., muscle stretch) followed by a slow ramping isometric contraction; they are off during rapid voluntary contraction [13].

The number **type II afferent** endings in a muscle spindle varies from zero to five, and they supply one intrafusal fiber terminating mainly on nuclear chain fibers. The endings of the secondary afferents are spirals ending on the polar ends of the intrafusal fibers (**Figure 1**). Secondary afferents have a regular tonic discharge, and do not exhibit an off-response at the termination of a voluntary ramp-and-hold contraction [42, 43].

Motor innervation. In addition to sensory neurons, intrafusal muscle fibers are also innervated by efferent motoneurons (fusimotor innervation). Axons of motoneurons enter the muscle spindle together with the sensory fibers and innervate intrafusal fibers in the polar regions forming motor endplates.

Motor innervation originates from myelinated **γ-motoneurons** (diameter 4–8 μm), also known as fusimotor motoneurons. They have been differentiated into static and dynamic. Dynamic axons have a weak effect on primary afferent firing while the static ones have a great influence on both primary and secondary afferents (see [43]).

Occasionally additional afferent innervation of muscle spindles originates from axons that also supply extrafusal muscle, known as **β-motoneurons** or skeleto-fusimotor fibers. These fibers supply both intrafusal and extrafusal fibers via motor endplates at the polar ends. The endplates of γ-motoneurons differ structurally from those formed by α-motoneurons on extrafusal fibers, but both are cholinergic synapses with many features in common, including junctional folds and a basal lamina filling the synaptic cleft. [42–46].

Stimulation of γ-motoneurons result in excitation of both Ia and II muscle spindle afferents. On the other hand, stimulation of **α-motoneurons** supplying extrafusal muscle fibers, results in coactivation of γ-motoneurons which in turn causes the contraction of the polar ends of the intrafusal fibers, restoring tension and sensitivity of the muscle spindle to stretch.

Thus, the γ-motoneuron function control the sensitivity of muscle spindle afferents as length detectors. Therefore, the muscle spindle's function as a length sensor arises essentially from its anatomical relationship with its parent muscle. Any length change in the parent muscle result in stretch of intrafusal fibers that is then detected by sensory receptors located on the equatorial and polar regions of the muscle spindle [44].

2.2 Golgi's tendon organ (tendon spindle)

The Golgi-tendon organ or tendon-spindle, localized at the origins and insertion of tendon, or rarely within the tendon. It is a mechanoreceptor that informs on muscle tension via its Ib afferent (**Figure 1**).

Structurally it consists of a capsule and within it there are loosely packed collagen fibers and muscle fibers (3–50). Among these elements there is a unique Ia afferent which branches to innervate the distal and the proximal parts of the organ [28, 47]. With respect to the skeletal muscle fibers the Golgi-tendon organ is in series between muscle and tendon.

The Golgi-tendon model react to "static and dynamic responses to activation of single motor units whose muscle fibers insert into the Golgi tendon organ, self and cross adaptation, non-linear summation when multiple motor units are active in the muscle, and the proportional relationship between the cross-adaptation and summation recorded for various pairs of motor units" [47, 48].

3. Mechanotransduction in muscle spindles

The sensory terminals of muscle spindles appear to adhere to the surface of the intrafusal muscle fibers, and although they possess a basal lamina in close contact with the plasmalemma it is absent at the sensory terminals.

3.1 Afferent glutamate-ergic neurotransmision in muscle spindles?

Bewick and co-workers [44] have demonstrate the occurrence of a complete gutamatergic neurotransmission system in the afferents of muscle spindles associated to the synaptic-like vesicles typical of those terminals. Exogenous glutamate enhances spindle excitability, an effect that can be pharmacologically blocked with specific molecules. On the other hand, synaptic-like vesicles contain glutamate, which is released during membrane cycling and, subsequently, a requirement for a replenishment mechanism.

This observation, however, does not exclude the possibility that other neuroactive substances also occur in these sensory terminals.

3.2 Ion channels and mechanorasduction in muscle spindles

In addition to the possible classical neurotransmission, the primary mechanism of mechanical transduction in muscle spindle sensory endings is the activation of stretch-sensitive ion channels. In mechanotransduction, i.e. the conversion of mechanical stimuli into biological or electrical signal, is triggered by members of the superfamilies of degenerin-epithelial Na^+-channels (Deg-ENa^+C; including acid-sensing ion channels -ASIC-), transient receptor potential channels (TRP), two-pore domain potassium (K_{2p}), and PIEZO [49, 50]. Some of them have been detected directly in proprioceptors as well as in primary sensory neurons innervating them. However, and similarly as in cutaneous mechanoreceptors, the stretch-sensitive channels responsible for transducing mechanical stimuli in spindle afferents awaits definitive identification (see [51]).

There is mounting evidence for the involvement of members of the Deg/ ENa^+C superfamily as mechanosensory channel(s) in mammalian primary afferent neurons, and in the sensory endings of muscle spindles [52–54]. All four subunits of the ENaC channel (α, β, γ and δ) are present in in spindle primary-sensory terminals [44, 54].

ASICs are members are a family of voltage-insensitive cation channels expressed in the nervous system and many types non-nervous cells. In rodents and humans six ASIC subtypes (ASIC1a, ASIC1b, ASIC2a, ASIC2b, ASIC3, and ASIC4) have been identified and their expression patterns are now rather well known [55, 56]. Regarding muscle spindles, evidence has been obtained in favor of a role of ASIC2 as primary mechanotransducer [53, 54]. Consistently, mice deficient in ASIC2, and also in ASIC3, show deficits in mechanical sensitivity [57–59].

PIEZO are Ca^{2+}-permeable mechanosensitive channels characterized by their large size and structure [60–62]. Piezo2 is expressed in proprioceptive dorsal root ganglia (DRG) neurons [63, 64] as well as sensory endings of proprioceptors innervating muscle spindles and Golgi tendon organs in mice [64]. Loss of PIEZO2 in proprioceptive neurons results in ataxia and dysmetria, severely uncoordinated body movements and abnormal limb positions, contracture of multiple joints, and muscle weakness, suggesting that PIEZO2 requirement for the activity of these mechanosensors [64–69]. Recently, an elegant study by the Ana Gomis's group corroborated these findings using mesencephalic nucleus proprioceptive neurons [70].

Regarding TRP channels there is little evidence for a role in low-threshold sensation in spindles.

4. Proprioceptive pathways

To drive proprioception to the central nervous system two different pathways must be considered: the unconscious proprioception is convoyed primarily via the spinocerebellar tracts to the cerebellum while the conscious proprioception is convoyed by the dorsal column-medial lemniscus pathway and the thalamus to the cerebral cortex.

Classically, the proprioceptive pathways of the spinal nerves have been described as a 2 neurons chain (**Figure 2**). The primer order neuron is a pseudo-unipolar neuron whose bodies are localized in DRG whose peripheral axonal branches reach proprioceptors (especially muscle spindles and Golgi tendon organs) and the central branch reach the base of the dorsal horn of the spinal cord. The second order neurons are placed in the medial Stilling-Clarke's column (which extends between the medullary segments C8 and L2) and the lateral Bechterew's column, both corresponding to the Rexed's lamina VII of the dorsal horn; in these columns the spinocerebellar tracts (dorsal or Foville-Flechsig fascicle and ventral or Gowers fascicle) originate to ascend and reach the cerebellum. This information is necessary unconscious. The spinocerebellar neurons together provide the major direct sensory projection from the hindlimbs and lower part of the trunk to the cerebellum. A parallel system serving the forelimbs includes the direct cuneocerebellar and rostral spinocerebellar tracts and other indirect pathways via the lateral reticular nucleus and the inferior olive.

Nevertheless, some aspects of the proprioception are conscious, and therefore the information must reach the cerebral cortex. For these components of the proprioceptive sensitivity, the proprioceptive pathways consist of a 3 neurons chain (**Figure 2**). The primary order neurons are placed in DRG and the central branch of their axons ascend throughout the dorsal columns of the spinal cord to reach the gracile and cuneate nuclei in the medulla. In those nuclei are placed the bodies of the secondary order neurons whose axons project to the ventral postero-lateral (VPL) nucleus of the thalamus (tertiary order neuron) whose axons end in the somatosensory cortex to provide the conscious perception of proprioception.

A particular question arises from cephalic muscles. They are innervated by cranial nerves and most of them (with the exception of jaw muscles and extraocular muscles) lack typical proprioceptors, i.e. muscle spindles. At present is commonly accepted that the proprioception of the cephalic territory depends on the trigeminal nerve [27, 71]. But the Gasser's ganglion of the trigeminal nerve does not contain proprioceptive neurons, while they are localized in the trigeminal mesencephalic nucleus.

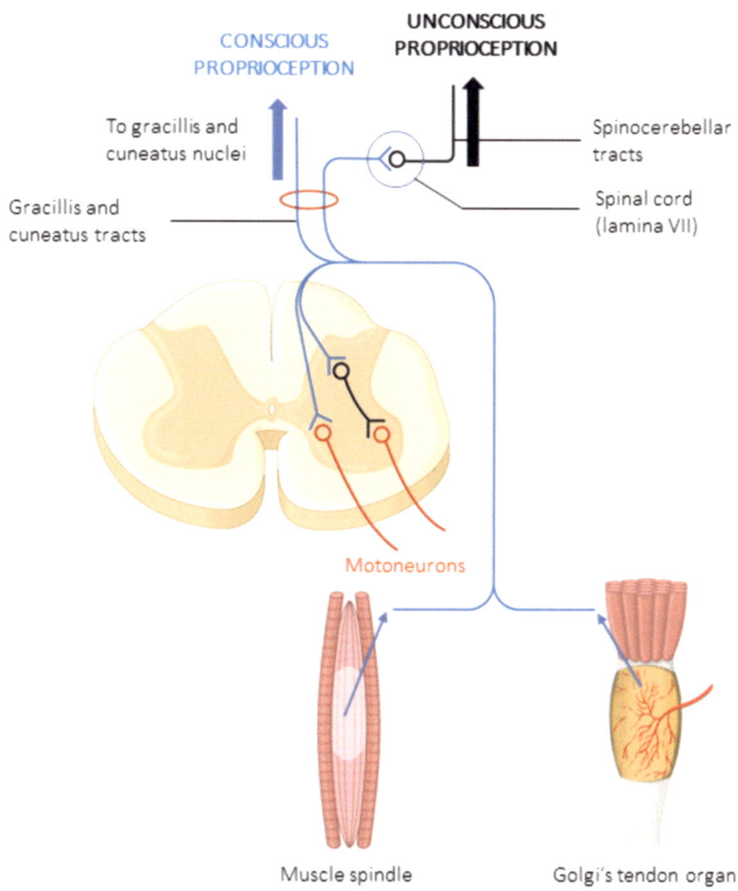

Figure 2.
Schematic representation of proprioceptive unconscious (black) and conscious (blue) pathways reaching the cerebellum and the brain, respectively.

4.1 Primary order neurons: the proprioceptive neurons

The proprioceptive neurons represent a small population (about 7–10%) of DRG primary sensory neurons and correspond with those with the largest cell bodies [72]. They can be classified and distinguished from other dorsal root ganglion neurons as a unique neuronal population using single cell transcriptome analysis [73]. They typically express the neurotrophin receptor TrkC (the preferred ligand for neurotrophin-3) and the Ca^{2+}-binding protein parvalbumin [74], as well as other markers that define this neuronal population [75].

The peripheral branch of the axons of the proprioceptive pseudo-unipolar neurons forms large-myelinated Aα and Aβ fibers in peripheral nerves, while the central branch stablish synapse in the spinal cord or ascends throughout the dorsal columns of the spinal cord to reach the gracile and cuneate nuclei in the medulla. Some peripheral branches, however, travel through the cuneatus tract and ascend the cervical spinal cord to reach the medulla oblongata to reach the accessory cuneate nucleus.

4.2 Secondary order neurons: the medial stilling-Clarke's (*nucleus dorsalis*) and lateral Bechterew's columns

The secondary order neurons of the proprioceptive pathways are localized at the base of the dorsal horn of the spinal cord, in correspondence with the VII Rexed's lamina, also known as the intermediate zone of the spinal cord. In this place two nuclear columns were classically considered: the nucleus dorsalis of Clarke (also known as Clarke's column, dorsal nucleus, posterior thoracic nucleus) and the medial nucleus (also known as Bechterew's nucleus). The axons of the neurons localized at these places form the spinocerebellar tracts.

4.3 The spinocerebellar tracts – unconscious proprioception

The tracts that carry unconscious proprioceptive information are collectively known as the spinocerebellar tracts. Within the spinocerebellar tracts, there are four individual pathways: ventral, dorsal, rostral and cuneocerebellar.

4.3.1 Ventral spinocerebellar tract (VSCT; anterior spinocerebellar, Gowers' fascicle)

VSCT originates mainly from the medial part of lamina 7 in the lumbosacral segments, from the dorsolateral nucleus of lamina 9 at L3-L6, and also from the neurons of the ventrolateral nucleus of lamina 9 and the lateral part of lamina 7 at L4-L5 segments [76, 77]. Axons decussate in the anterior white commissure and run in the ventral border of the lateral funiculi. They ascend through the brainstem to the pons where turn dorsally and enter the cerebellum throughout the superior cerebellar peduncle. For the most part, the ventral spinocerebellar tract axons recross the midline in the deep white matter of the cerebellum to terminate ipsilaterally. It terminates in the cortex of the anterior lobe and vermis of the posterior lobe and give collaterals to the globose and emboliform nuclei.

4.3.2 Dorsal spinocerebellar tract (DSCT; posterior spinocerebellar tract, Flechsig's fasciculus, Foville-Flechsig fasciculus)

This nucleus is present from Th1 through the second lumbar spinal segments and is largest in the lower thoracic and upper lumbar segments [77]. The cells of origin of the DSCT are classically described as residing in Clarke's column of the lumbar and thoracic spinal cord segments. The *nucleus dorsalis* of Clarke is a group of neurons localized in the medial part of lamina VII extending from C7 to L2 levels related to the proprioception of the lower limb. Caudally it begins at L2 level and reaches its maximum at Th12 level and above Th8 level its size diminishes, and the column ends at C7 level. Nevertheless, it is represented in other spinal regions by scattered neurons, which become aggregated to form a cervical nucleus at C3 level and a sacral nucleus in the middle of the sacral spine region (**Figure 2**).

In addition, other groups of neurons that also belong to DSCT are located throughout the intermediate and dorsal laminae of the thoracic and lumbosacral segments of the spinal cord. The axons of the DSCT ascends ipsilaterally in the peripheral region of the funiculus lateralis of the spinal cord. Then, they continue to course through the medulla oblongata and finally pass through the inferior cerebellar peduncle and into the cerebellum, and terminate in the cerebellar cortex of lobules I–V, in the anterior lobe and in the posterior lobe vermis and paramedian lobe. In addition to those cortical projections, there is evidence that DSCT fibers also terminate in the medial and interpositus cerebellar nuclei.

4.3.3 Rostral spinocerebellar tract

The rostral spinocerebellar tract appears to be the upper extremity homolog of the ventral spinocerebellar tract. The neurons of origin of this tract are located in Rexed laminae V-VII of the C5-C8 segments. The projection is predominantly ipsilateral, but there is also a minor bilateral projection. The axons of the rostral spinocerebellar tract neurons terminate in the anterior and paramedian lobule cerebellar.

4.3.4 Cuneocerebellar tract (posterior external arcuate fibers, dorsal external arcuate fibers)

DSCT does not convey information from the upper limb since the *nucleus dorsalis* does not extend into the cervical spinal cord. Therefore, there is another proprioceptive pathway for the upper limb: the cuneocerebellar tract. The secondary order neurons of this nucleus pass to the inferior ipsilateral cerebellar peduncle to reach the spinocerebellum.

4.3.5 Other spinocerebellar tracts

There are additional ascending direct and indirect spinocerebellar pathways. The *spinocervical tract* relays in the lateral cervical nucleus and projects to the ventral postero-lateral (VPL) nucleus of the thalamus. The *spinohypothalamic* and *spinoamygdalar* tract provides sensory input to areas of the nervous system involved in controlling autonomic, endocrine and emotional responses. Sensory information from the hindlimbs is also relayed by indirect *spinoreticulocerebellar* pathways through at least two olivocerebellar pathways: the indirect *spinoreticuloolivocerebellar* tract and the direct *spinoolivocerebellar* tract.

These pathways provide necessary information regarding the current status of reflex pathways, as well as muscle tone, length and tension that consent the cerebellum to coordinate and regulate motor activity.

4.4 The brain connection – conscious proprioception

According to Proske and Gadivia [4] there are at least two reasons for including body schemas and images in the study and discussion of proprioception. "*First, while proprioceptors provide information about position and movement of the limb, they are unable to signal the length of limb segments and therefore the absolute location of the limb in space. Second, there is the issue of body ownership*". The blind movement of a limb, while proprioceptive feedback informs us about the movement, we need to be able to identify the moving limb as our own [78, 79]. Carruthers has proposed that all representations of the body are available to consciousness [80]. "On-line," newly constructed body representations, provided by inputs from vision, touch, and proprioception, generate a perception of the body as it actually is at any moment in time, an image which is able to change from moment to moment. It is distinguished from an "off-line" representation constructed, in part, from current sensory inputs, in part, from stored memories and is available to consciousness both immediately and after retrieval of memories.

In the last times neuroimaging has strongly contributed to the knowledge of the central activity patterns produced by proprioceptive stimuli, to the recognition of the integration of proprioceptive inputs with inputs from other senses and the identification of central areas involved in the integration [81]. These include regions in the parietal cortex [82–85] including the primary somatosensory cortex [86, 87]. Furthermore, parts of the frontal cortex and insula [88, 89] are involved in proprioception.

4.5 The mesencephalic trigeminal nucleus

The mesencephalic trigeminal nucleus is a sensory structure located at the mesopontine junction and contains the cell bodies of primary order afferent proprioceptors that innervate muscle spindles of the muscles of mastication and other muscles of the head and neck [27, 71]. Whether these primary sensory neurons are generated directly in brain or have a of neural crest origin is stell debated. Classically it has been regarded as a representation of a peripheral sensory ganglion similar to DRG that became incorporated into de brainstem during embryonic development, although molecular studies support a central origin for these cells [90, 91].

It projects to the dorsolateral division of the trigeminal motor nucleus and to the supratrigeminal nucleus, which are involved in humans in the jaw-jerk reflex and the periodontal-masseteric reflex [92] (for a review see [93]).

5. Propioception from a clinical perspective: causes for impaired proprioception

Proprioception is critical factor for stability, and it is well known it deteriorates in aged people [94]. In fact, the proprioceptive system undergoes significant structural and functional changes with aging which cause a progressive decline in somatosensibility including proprioception [95–97].

Aging courses with muscle weakness from sarcopenia, decrease in the number of intrafusal fibers in muscle spindles, and denervation. All together these facts diminished muscle force and consistently the proprioception which, in turn determinate the increase in falls in the elderly with clinical and public health consequences. The specific relationship between muscle strength and proprioception should be explored further as it may provide a basis for the claim that exercise improves standing stability. Interestingly, to achieve a reduction in the incidence of falls, it is not sufficient to improve muscle strength alone as exercises are required which actually challenge standing stability. These changes might contribute to the frequent falls and motor control problems observed in older adults. On the structural level, muscle spindles in aged humans possess fewer intrafusal fibers, an increased capsular thickness and some spindles which show signs of denervation [98, 99].

A variety of neurological diseases are characterized by irregular, jerky movement or posture due to loss of proprioceptive sensory feedback, a disturbance called afferent ataxia. The affected neurons are primary sensory neurons in the dorsal root ganglia relaying body position and movement (proprioception) to the central nervous system. Proprioception dysfunction can be caused by injuries and disorders that affect any part of the proprioceptive system between the sensory receptors that send the signals to the parts of the brain that receive and interpret them. The risk of proprioception loss increases as we age due to a combination of natural age-related changes to the nerves, joints, and muscles. Examples of injuries and conditions that can cause proprioceptive deficit include: brain injuries, herniated disc, arthritis, multiple sclerosis (MS), stroke, autism spectrum disorder (ASD), diabetes, peripheral neuropathy, Parkinson's disease, Huntington's disease, ALS (amyotrophic lateral sclerosis), or Lou Gehrig's disease. Joint injuries, such as an ankle sprain or knee sprain, joint replacement surgery, such as hip replacement or knee replacement, Parkinson's disease. Primary proprioceptive neurons may be the target of hereditary, developmental, degenerative, toxic, inflammatory and autoimmune pathology. Accordingly, typical clinical consequences of pathology affecting proprioceptive neurons, in addition to afferent ataxia, include loss of deep tendon (stretch) reflexes and of conscious perception of position and movement of body parts, often associated with loss of perception of vibration [9]. Recently, Dionisi and co-workers have obtained proprioceptive primary sensory neurons from iPSCs, and the generation of intrafusal fibers in vitro are opening new perspectives for the treatment of some ataxia linked to altered primary proprioceptive neurons [100, 101].

Author details

José A. Vega[1,2]* and Juan Cobo[3,4]

1 Departamento de Morfología y Biología Celular, Grupo SINPOS, Universidad de Oviedo, Oviedo, Spain

2 Facultad de Ciencias de la Salud, Universidad Autónoma de Chile, Santiago de Chile, Chile

3 Departamento de Cirugía y Especialidades Médico-Quirúrgicas, Universidad de Oviedo, Oviedo, Spain

4 Instituto Asturiano de Odontología, Oviedo, Spain

*Address all correspondence to: javega@uniovi.es

References

[1] Kröger S, Watkins B. Muscle spindle function in healthy and diseased muscle. Skelet Muscle. 2021; 11:3.

[2] Williams SR, Chapman CE. Time course and magnitude of movement-related gating of tactile detection in humans. II. Effects of stimulus intensity. J Neurophysiol. 2000; 84: 863-875.

[3] Ciancia F, Maitte M, Coquery JM. Reduction during movement of the evoked potentials recorded along the extralemniscal pathways of the cat. Electroencephalogr Clin Neurophysiol. 1980; 48: 197-202.

[4] Proske U, Gandevia SC. The kinaesthetic senses. J Physiol. 2009; 587: 4139-4146.

[5] Proske U, Gandevia SC. The proprioceptive senses: their roles in signaling body shape, body position and movement, and muscle force. Physiol Rev. 2012; 92:1651-1697.

[6] Ferlinc A, Fabiani E, Velnar T, Gradisnik L. The Importance and Role of Proprioception in the Elderly: a Short Review.Mater Sociomed. 2019; 31: 219-221.

[7] Henry M, Baudry S. Age-related changes in leg proprioception: implications for postural control. J Neurophysiol. 2019; 122: 525-538.

[8] Ganz DA, Latham NK1. Prevention of Falls in Community-Dwelling Older Adults. N Engl J Med. 2020; 382: 734-743.

[9] Pandolfo M, Manto M. Cerebellar and Afferent Ataxias. Continuum Lifelong Learn Neurology 2013; 19: 1343.

[10] Klockgether T, Mariotti C, Paulson HL. Spinocerebellar ataxia. Nat Rev Dis Primers. 2019; 5: 24.

[11] Proske U, Gandevia SC. Kinesthetic Senses. Compr Physiol. 2018; 8: 1157-1183.

[12] Proske U. Exercise, fatigue and proprioception: a retrospective. Exp Brain Res. 2019; 237: 2447-2459.

[13] Banks RW. The innervation of the muscle spindle: a personal history. J Anat. 2015; 227: 115-135.

[14] Kröger S. Proprioception 2.0: novel functions for muscle spindles. Curr Opin Neurol. 2018; 31:592-598.

[15] Grigg P. Peripheral neural mechanisms in proprioception. J Sport Rehab. 1994; 3: 2-17.

[16] Banks RW, Hulliger M, Saed HH, Stacey MJ. A comparative analysis of the encapsulated end-organs of mammalian skeletal muscles and of their sensory nerve endings. J Anat. 2009; 214: 859-887.

[17] Burke D, Gandevia SC, Macefield G. Responses to passive movement of receptors in joint, skin and muscle of the human hand. J Physiol. 1988; 402: 347-361.

[18] Collins DF, Prochazka A. Movement illusions evoked by ensemble cutaneous input from the dorsum of the human hand. J Physiol. 1996; 496: 857-871.

[19] Frith C. The self in action: lessons from delusions of control. Consciousness Cognition. 2005; 14: 752-770.

[20] Johansson RS, Flanagan JR. Coding and use of tactile signals from the fingertips in object manipulation tasks. Nat Rev Neurosci. 2009; 10: 345-359.

[21] Zimmerman A, Bai L, Ginty DD. The gentle touch receptors

of mammalian skin. Science. 2014; 346: 950-954.

[22] Cobo R, García-Piqueras J, Cobo J, Vega JA. The Human Cutaneous Sensory Corpuscles: An Update. J Clin Med. 2021; 10(2). pii: E227.

[23] Gandevia SC, Miller S, Aniss AM, Burke D. Reflex influences on muscle spindle activity in relaxed human leg muscles. J Neurophysiol. 1986; 56: 159-170.

[24] Edin BB. Quantitative analyses of dynamic strain sensitivity in human skin mechanoreceptors. J Neurophysiol. 2004; 92: 3233-3243.

[25] Izumizaki M, Tsuge M, Akai L, Proske U, Homma I. The illusion of changed position and movement from vibrating one arm is altered by vision or movement of the other arm. J Physiol. 2010; 588: 2789-2800.

[26] Cobo JL, Abbate F, de Vicente JC, Cobo J, Vega JA. Searching for proprioceptors in human facial muscles. Neurosci Lett. 2017; 640:1-5.

[27] Cobo JL, Solé-Magdalena A, Junquera S, Cobo T, Vega JA, Cobo J. The Proprioception in the Muscles Supplied by the Facial Nerve. In: *Selected Topics in Facial Nerve Disorders*, I. Al-Zwaini and M.J Hussein Eds. InTechOpen, London, 2019, pp. 15.28.

[28] Jami L. Golgi tendon organs in mammalian skeletal muscle: functional properties and central actions. Physiol Rev. 1992; 72: 623-666.

[29] Proske U. The tendon organ. In: Peripheral Neuropathy, edited by Dyck P, Thomas P, Griffin J, Low P, Poduslo J. Philadelphia, PA: Saunders, 1993, p. 141-148.

[30] Banks RW. The muscle spindle. In: Dyck PJ, Thomas PK, editors. Peripheral neuropathy. 4. Philadelphia: WB Saunders; 2005. pp. 131-150.

[31] Banks RW. The innervation of the muscle spindle: a personal history. J Anat. 2015; 227:115-135.

[32] Soukup T, Pederosa-Domellöf F, Thornell LE: Intrafusal fiber type composition of muscle spindles in the first human lumbrical muscle, Acta Neuropathol (Berl). 2003; 105:18-24.

[33] Kulkarni V, Chandy MJ, Babu KS: Quantitative study of muscle spindles in suboccipital muscles of human foetuses. Neurol India. 2001; 49: 355-359.

[34] Liu J-X, Eriksson PO, Thornell L-E, Pedrosa-Domellof F. Myosin heavy chain composition of muscle spindles in human biceps brachii, J Histochem Cytochem. 2002; 50:171-184.

[35] Liu J-X, Eriksson PO, Thornell L-E, Pedrosa-Domellof F. Fiber content and myosin heavy chain composition of muscle spindles in aged human biceps brachii. J Histochem Cytochem. 2005; 53: 445-454.

[36] Banks RW, Harker DW, Stacey MJ. A study of mammalian intrafusal muscle fibres using a combined histochemical and ultrastructural technique. J Anat. 1977; 123: 783-796.

[37] Liu J-X, Thornell L-E, Pedrosa-Domellof F. Muscle spindles in the deep muscles of the human neck: a morphological and immunocytochemical study. J Histochem Cytochem. 2003; 51: 175-186.

[38] Ovalle WK, Smith RS: Histochemical identification of three types of intrafusal muscle fibers in cat and monkey based on myosin ATPase reaction Can J Physiol Pharmacol. 1972; 50:195-202.

[39] Kucera J, Dorovini-Zis K. Types of human intrafusal muscle fibers. Muscle Nerve. 1979; 2:437-451.

[40] Thornell LE, Carlsson L, Eriksson PO, Liu JX, Österlund C, Stål P, Pedrosa-Domellöf F. Fibre typing of intrafusal fibres. J Anat. 2015; 227: 136-156.

[41] Matthews PB. Where Anatomy led, Physiology followed: a survey of our developing understanding of the muscle spindle, what it does and how it works. J Anat. 2015;227: 104-114.

[42] Edin BB, Vallbo AB. Dynamic response of human muscle spindle afferents to stretch. J Neurophysiol. 1990a; 63: 1297-1306.

[43] Proske U. The mammalian muscle spindle. News Physiol Sci. 1997; 12: 37-42.

[44] Bewick GS, Banks RW. Mechanotransduction in the muscle spindle. Pflugers Arch. 2015; 467:175-90.

[45] Banks RW. The motor innervation of mammalian muscle-spindles. Prog Neurobiol. 1994;43: 323-362.

[46] Manuel M, Zytnicki *D. alpha*, beta and gamma motoneurons: functional diversity in the motor system's final pathway. J Integr Neurosci. 2011; 10: 243-276.

[47] Proske U. The Golgi tendon organ, Int Rev Physiol. 1981; 25:127-171.

[48] Mileusnic MP, Loeb GE. Mathematical models of proprioreceptors. II Structure and function of the Golgi tendon organ, J Neurophysiol. 2006; 96:1789-1802.

[49] Delmas P, Coste B. Mechano-gated ion channels in sensory systems. Cell. 2013; 155:278-284.

[50] Cobo R, García-Piqueras J, García-Mesa Y, Feito J, García-Suárez O, Vega JA. Peripheral Mechanobiology of Touch-Studies on Vertebrate Cutaneous Sensory Corpuscles. Int J Mol Sci. 2020; 21(17). pii: E6221.

[51] Anderson EO, Schneider ER, Bagriantsev SN. Piezo2 in Cutaneous and Proprioceptive Mechanotransduction in Vertebrates. Curr Top Membr. 2017; 79: 197-217.

[52] Ismailov II, Berdiev BK, Shlyonsky VG, Benos DJ. Mechanosensitivity of an epithelial Na+ channel in planar lipid bilayers: release from Ca^{2+} block. Biophys J. 1997; 72:1182-1192.

[53] Simon A, Shenton F, Hunter I, BanksRW,BewickGS.Amiloride-sensitive channels are a major contributor to mechanotransduction in mammalian muscle spindles, J Physiol. 2010; 588: 171-85.

[54] Althaus M, Bogdan R, Clauss WG, Fronius M. Mechano-sensitivity of epithelial sodium channels (ENaCs): laminar shear stress increases ion channel open probability. FASEB J. 2007; 21:2389-2399.

[55] Baron A, Lingueglia E. Pharmacology of acid-sensing ion channels - physiological and therapeutical perspectives. Neuropharmacology. 2015; 94:19-35. Cheng YR, BY, Chen CC. Acid-sensing ion channels: dual function proteins for chemo-sensing and mechano-sensing. J Biomed Sci. 2018; 25: 46.

[56] Lee CH, Chen CC. Roles of ASICs in Nociception and Proprioception. Adv Exp Med Biol. 2018; 1099: 37-47.

[57] Chen CC, Wong CW. Neurosensory mechanotransduction through acid-sensing ion channels. J Cell Mol Med. 2013 ;17: 337-49.

[58] Lin SH, Sun WH, Chen CC. Genetic exploration of the role of acid-sensing ion channels. Neuropharmacology. 2015; 94: 99-118.

[59] Omerbasic D, Schuhmacher LN, Bernal Sierra YA, Smith ES. Lewin GR. ASICs and mammalian mechanoreceptor function. Neuropharmacology. 2015; 94:80-86.

[60] Murthy SE, Dubin AE, Patapoutian A. Piezos thrive under pressure: mechanically activated ion channels in health and disease. Nat Rev Mol Cell Biol. 2017; 18:771-83.

[61] Saotome K, Murthy SE, Kefauver JM, Whitwam T, Patapoutian A, Ward AB. Structure of the mechanically activated ion channel Piezo1. Nature. 2017; 554: 481.

[62] Zhao Q, Zhou H, Chi S, Wang Y, Wang J, Geng J, Wu K, Liu W, Zhang T, Dong MQ, Wang J, Li X, Xiao B.Structure and mechanogating mechanism of the Piezo1 channel. Nature. 2018; 554: 487-492.

[63] Woo SH, Lukacs V, de Nooij JC, Zaytseva D, Criddle CR, Francisco A, Jessell TM, Wilkinson KA, Patapoutian A. Piezo2 is the principal mechanotransduction channel for proprioception. Nat Neurosci. 2015; 18:1756-1862.

[64] Woo SH, Ranade S, Weyer AD, Dubin AE, Baba Y Qiu Z, Petrus M, Miyamoto T, Reddy K, Lumpkin EA, Stucky CL, Patapoutian A. Piezo2 is required for Merkel-cell mechanotransduction. Nature. 2014; 509:622-6.

[65] Chesler AT, Szczot M, Bharucha-Goebel D, Čeko M, Donkervoort S, Laubacher C, Hayes LH, Alter K, Zampieri C, Stanley C, Innes AM, Mah JK, Grosmann CM, Bradley N, Nguyen D, Foley AR, Le Pichon CE, Bönnemann CG. The role of *PIEZO2* in human mechanosensation. N Engl J Med. 2016; 375: 1355-64.

[66] Haliloglu G, Becker K, Temucin C, Talim B, Küçükşahin N, Pergande M, Motameny S, Nürnberg P, Aydingoz U, Topaloglu H, Cirak S. Recessive PIEZO2 stop mutation causes distal arthrogryposis with distal muscle weakness, scoliosis and proprioception defects. J Hum Genet. 2017; 62:597-501.

[67] Delle Vedove A, Storbeck M, Heller R, Hölker I, Hebbar M, Shukla A, Magnusson O, Cirak S, Girisha KM, O'Driscoll M, Loeys B, Wirth B. Biallelic loss of proprioception-related PIEZO2 causes muscular atrophy with perinatal respiratory distress, arthrogryposis, and scoliosis. Am J Hum Genet. 2016; 99:1206-16.

[68] Mahmud AA, Nahid NA, Nassif C, Sayeed MS, Ahmed MU, Parveen M, Khalil MI, Islam MM, Nahar, Z, Rypens F, Hamdan FF, Rouleau GA, Hasnat A, Michaud JL. Loss of the proprioception and touch sensation channel PIEZO2 in siblings with a progressive form of contractures. Clin Genet. 2017; 91:470-5.

[69] Assaraf E, Blecher R, Heinemann-Yerushalmi L, Krief S, Carmel Vinestock R, Biton IE, Brumfeld V, Rotkopf R, Avisar E, Agar G, Zelzer E. Piezo2 expressed in proprioceptive neurons is essential for skeletal integrity. Nat Commun. 2020; 11: 3168.

[70] Florez-Paz D, Bali KK, Kuner R, Gomis A. A critical role for Piezo2 channels in the mechanotransduction of mouse proprioceptive neurons. Sci Rep. 2016; 6:25923.

[71] Lazarov NE. Neurobiology of orofacial proprioception. Brain Research Reviews. 2007; 56:362-383.

[72] Marmigère F, Ernfors P. Specification and connectivity of neuronal subtypes in the sensory lineage. Nat Rev Neurosci. 2007; 8: 114-127.

[73] Oliver KM, Florez-Paz DM, Badea TC, Mentis GZ, Menon V, de Nooij JC. Molecular development of muscle spindle and Golgi tendon organ sensory afferents revealed by single proprioceptor transcriptome analysis. bioRxiv. 2020. DOI: 10.1101/2020.04.03.023986.

[74] Hippenmeyer S. Vrieseling E, Sigrist M, Portmann T, Laengle C, Ladle DR, Arber S.. A Developmental Switch in the Response of DRG Neurons to ETS Transcription Factor Signaling. Plos Biol. 2005; 3: e159.

[75] Poliak S, Norovich AL, Yamagata M, Sanes JR, Jessell TM. Muscle-type Identity of Proprioceptors Specified by Spatially Restricted Signals from Limb Mesenchyme. Cell 2016; 164: 512-525.

[76] Xu Q, Grant G. Course of spinocerebellar axons in the ventral and lateral funiculi of the spinal cord with projections to the posterior cerebellar termination area: an experimental anatomical study in the cat, using a retrograde tracing technique. Exp Brain Res. 2005; 162:250-256.

[77] Watson C, Paxinos G, Kayalioglu G (Eds). *The spinal cord.* Academic Press, New York, 2008.

[78] De Vignemont F. Body schema and body image–pros and cons. Neuropsychologia 2010; 48: 669-680.

[79] Longo MR, Haggard P. An implicit body representation underlying human position sense. Proc Natl Acad Sci USA. 2010; 107: 11727-11732.

[80] Carruthers G. Types of body representation and the sense of embodiment. Conscious Cogn. 2008; 17: 1302-1316.

[81] Sanchez-Vives MV, Slater M. From presence to consciousness through virtual reality. Nat Rev Neurosci. 2005; 6: 332-339.

[82] Ehrsson HH, Holmes NP, Passingham RE. Touching a rubber hand: feeling of body ownership is associated with activity in multisensory brain areas. J Neurosci, 2005; 25: 10564-10573.

[83] Ehrsson HH, Spence C, Passingham RE. That's my hand! Activity in premotor cortex reflects feeling of ownership of a limb. Science 2004; 305: 875-877.

[84] Pellijeff A, Bonilha L, Morgan PS, McKenzie K, Jackson SR. Parietal updating of limb posture: an event-related fMRI study. Neuropsychologia. 2006; 44: 2685-2690.

[85] Hagura N, Oouchida Y, Aramaki Y, Okada T, Matsumura M, Sadato N, Naito E. Visuokinesthetic perception of hand movement is mediated by cerebro-cerebellar interaction between the left cerebellum and right parietal cortex. Cereb Cortex. 2009; 19: 176-186.

[86] Schaefer M, Flor H, Heinze HJ, Rotte M. Morphing the body: illusory feeling of an elongated arm affects somatosensory homunculus. Neuroimage. 2007; 36: 700-705.

[87] Schaefer M, Heinze HJ, Rotte M. My third arm: shifts in topography of the somatosensory homunculus predict feeling of an artificial supernumerary arm. Hum Brain Mapp. 2009; 30: 1413-1420.

[88] Schwoebel J, Coslett HB. Evidence for multiple, distinct representations of the human body. J Cogn Neurosci. 2005; 17: 543-553.

[89] Tsakiris M, Hesse MD, Boy C, Haggard P, Fink GR. Neural signatures of body ownership: a sensory network for bodily self-consciousness. Cereb Cortex. 2007; 17: 2235-2244.

[90] Dyer C, Linker C, Graham A, Knight R. Specification of sensory neurons occurs through diverse developmental programs functioning in the brain and spinal cord. Dev Dyn. 2014; 243: 1429-1439.

[91] Müller F, O'Rahilly R. The initial appearance of the cranial nerves and related neuronal migration in staged human embryos. Cells Tissues Organs. 2011; 193: 215-238.

[92] Shigenaga Y, Yoshida A, Mitsuhiro Y, Doe K, Suemune S. Morphology of single mesencephalic trigeminal neurons innervating periodontal ligament of the cat. Brain Res. 1988: 448: 331-338.

[93] Yoshida A, Moritani M, Nagase Y, Bae YC. Projection and synaptic connectivity of trigeminal mesencephalic nucleus neurons controlling jaw reflexes. J Oral Sci. 2017; 59: 177-182.

[94] Ribeiro F, Oliveira J. Aging effects on joint proprioception: the role of physical activity in proprioception preservation. Eur Rev Aging Phys Act. 2007; 4: 71-76.

[95] Verschueren SM, Brumagne S, Swinnen SP, Cordo PJ. The effect of aging on dynamic position sense at the ankle. Behav Brain Res. 2002; 136: 593-603.

[96] Landelle C, El Ahmadi A, Kavounoudias A. Age-related impairment of hand movement perception based on muscle proprioception and touch. Neuroscience. 2018; 381:91-104.

[97] García-Piqueras J, García-Mesa Y, Cárcaba L, Feito J, Torres-Parejo I, Martín-Biedma B, Cobo J, García-Suárez O, Vega JA. Ageing of the somatosensory system at the periphery: age-related changes in cutaneous mechanoreceptors. J Anat. 2019; 234:839-852.

[98] Swash M, Fox KP. The effect of age on human skeletal muscle. Studies of the morphology and innervation of muscle spindles. J Neurol Sci. 1972; 16: 417-432.

[99] Liu JX, Eriksson PO, Thornell LE, Pedrosa-Domellof F. Fiber content and myosin heavy chain composition of muscle spindles in aged human biceps brachii. J Histochem Cytochem. 2005; 53: 445-454.

[100] Dionisi C, Rai M, Chazalon M, Schiffmann SN, Pandolfo M. Primary proprioceptive neurons from human induced pluripotent stem cells: a cell model for afferent ataxias. Sci Rep. 2020; 10: 7752.

[101] Barret P, Quick TJ, Mudera V, Player DJ. Generating intrafusal skeletal muscle fibres in vitro: Current state of the art and future challenges. J Tissue Eng. 2020; 11: 1-15.

Chapter 2

Proprioception and Clinical Correlation

Pinar Gelener, Gözde İyigün and Ramadan Özmanevra

Abstract

Proprioception is the sense of position or the motion of the limbs and body in the absence of vision. It is a complex system having both conscious and unconscious components involving peripheral and central pathways. The complexity of sensorimotor systems requires deep knowledge of anatomy and physiology to analyze and localize the symptoms and the signs of the patients. Joint sense and vibration sense examination is an important component of physical examination. This chapter consists anatomy, motor control, postural control related to proprioception with neurologic clinical correlation and also the information about the changes of proprioception after orthopedic surgeries and discuss with the available literature.

Keywords: proprioception, neurology, orthopedics

1. Introduction

1.1 Anatomy

Proprioception was first described by Sir Charles Bell in 1830s as sixth sense coming from Latin word proprius meaning "one's own" and perception "perceiving one's own self" [1]. Proprioception is generally defined as either the sense of position or the motion of the limbs and body in the absence of vision [2]. Limb position is a static sense, whereas limb motion is a dynamic sense [3]. It is described as the most important sensorial modality for the internal representation of body map providing static and dynamic proprioceptive systems [4].

Proprioception is a complex system having both conscious and unconscious components involving peripheral and central pathways. The proprioceptive sensations arise from the deeper tissues. The main receptors are muscle spindles, tendons, Ruffini endings in joint capsules ligaments and Pacinian corpuscles reacting pressure, tension, stretching or contraction. The cutaneous receptors of the skin also contribute to joint position and motion sense especially at digits, elbow and knee. The term kinesthesia is generally used to describe the conscious awareness of the body or limb position in space [1, 5–7]. Conscious proprioceptive impulses elongate along large and myelinated fibers from the peripheral nerves into the dorsal root ganglion of spinal cord (first order neurons) and then via the medial division of the posterior root, via posterior white columns of fasciculi gracilis and cuneatus and ascend to the nuclei gracilis and cuneatus in the lower medulla. Axons of the second-order neuron decussate as internal arcuate fibers (second order neurons), and then ascend in the medial lemniscus to the contralateral somatosensory region of thalamus (**Figure 1**) [2, 5].

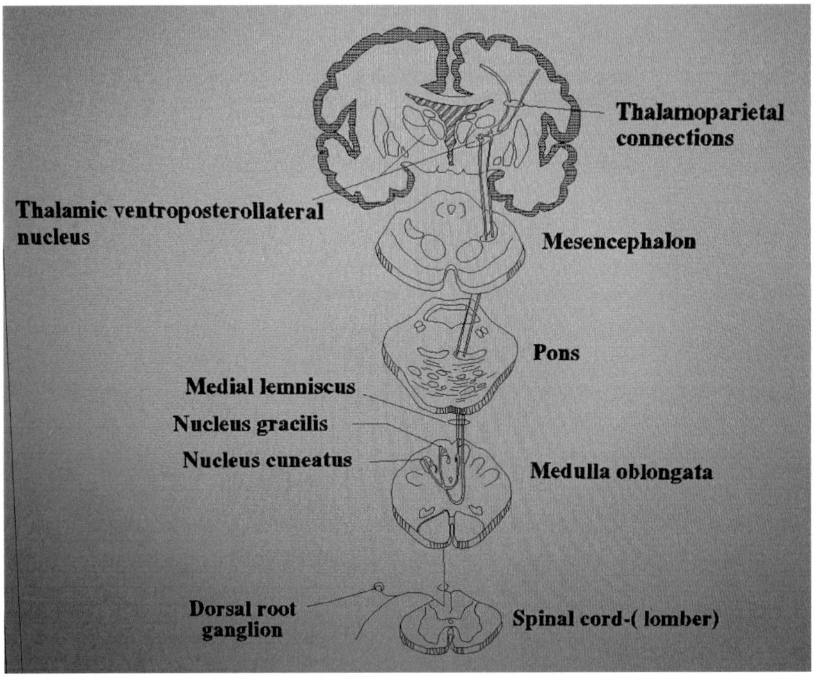

Figure 1.
Neuroanatomic pathway adopted from DeJong's the neurologic examination.

The main pathway for proprioceptive information is via the dorsal column medial lemniscal, posterior and anterior spinocerebellar tracts and spinoreticular tracts [6, 7].

There is a high density of complex spindles in deeper cervical muscles particularly in the intermediate columns, acting as neck prorioceptive receptors. This system is important for head and neck position sense together with the high density muscle spindles of sub-occipital triange. The density of muscle spindles is higher in the upper cervical spine when compared with the lower cervial and cervico-thoracic and thoraco-lumbar junctions [8]. Neck proprioception plays an important role in limb coordination and body-scheme representation [8]. Proprioceptive impulses from the head and neck are supplied by cranial nerves [5].

Contralateral primary and secondary sensorimotor cortex, supplementary motor area and bilateral inferior parietal lobes and basal ganglia (especially nigrostriatal pathways, striatal neurons and putamen) are involved in processing proprioceptive information during passive movement [9, 10]. The cerebellum contributes to proprioception only during movement [3]. Especially deep medial fastigial nucleus of cerebellum converges vestibular and neck proprioceptive sensory signals describing body's movement in space [11, 12].

1.2 Proprioception and motor control

The sensorimotor system, defined as the sensory, motor, and central integration, is a crucial and intricate component of the motor control system [13]. Motor control is a complex and dynamic process based on the selective integration of sensory information from multiple sources, motor commands, and motor output [13, 14]. There are specific unique roles associated with each sensory source (i.e., somatosensory, visual, vestibular) that cannot be compensated fully with each other [14, 15]. The environment is experienced through sensory systems: exteroception (e.g., sight, hearing, touch), interoception (e.g., arousal, pain, visceral sensations, muscular sensations), and proprioception (e.g., sense of position, motion, and force), which all required for successful motor control [16, 17]. During a task-oriented activity, motor adaptation, defined as a process of modifying the movement based on error feedback [18], skills are needed to cope with the changes occurring in the *external* and *internal environment* [2]. Motor adaptation is stimulated with sensorial triggers by using both feedback (reactive: adjust ongoing motor behavior) and feedforward (preparatory: pre-planning and anticipating the motor sequence from the previous experience) mechanism. Proprioceptive information, from proprioceptors found in muscle, tendon, ligament, capsule, skin, and fascial layers, plays an integral role in motor control and considered as multifold [14].

The role of proprioceptive information in motor control can be divided into two categories: *external environment* (even vs. uneven ground) and *internal environment* (carrying a load on shoulders vs. hands below knuckle height). The motor programs often need to be adjusted to accommodate unexpected perturbations or changes in the *external environment*. Although the source of this information is usually associated mainly with visual input, there are many situations where proprioceptive input is the fastest and/or most accurate. Proprioception is necessary during motion execution to update feedforward commands derived from the visual image [14]. Attention to environmental constraints is also required because dealing with complex environments often requires behavioral flexibility to maintain postural balance [19]. Secondly, the central nervous system needs an updated body schema of the biomechanical and spatial properties of body parts to plan and modify *internally generated* motor commands [20]. Before and during a motor command, the motor control system must consider the current and changing positions of the respective joints to account for the complex mechanical interactions within the musculoskeletal system components [14]. Additionally, proprioception is important after movement to compare the actual movement and intended movement, besides the predicted movement derived from the efference copies (corollary discharge: copying of motor commands based on past events) of motor commands, which has an essential role in motor learning to update the internal forward model of motor command [21].

1.3 Proprioception and postural control

During the execution of all motor tasks, proprioception is required to prepare, maintain, and restore the stability of both the entire body (postural equilibrium) and the segments (joint stability) [14, 15]. Postural control, defined as controlling the position of the body regarding the task in the environment, involves neural control of "postural equilibrium" and "postural orientation". Postural equilibrium consists of the coordination of sensory and motor strategies to maintain balance by controlling the body's center of mass (COM) over its base of support (BOS) to maintain postural stability during both intrinsic (self-initiated) and extrinsic (externally triggered) disturbances. The postural equilibrium controls stability during both static (i.e., quiet standing) and dynamic (i.e., walking and reaching) situations. Postural orientation involves positioning body alignment with respect to gravity, the support surface, visual environment, and other sensory reference frames [22].

Postural control is considered as a complex motor skill derived from the interaction of multiple sensorimotor processes, which are; biomechanical constraints (i.e., BOS, degrees of freedom, strength, limits of stability), movement strategies (i.e., reactive, anticipatory, voluntary), sensory strategies (i.e., sensory integration, sensory re-weighting), orientation in space (i.e., perception of visional verticality, perception of postural verticality), control of dynamics (i.e., gait, proactive), cognitive processing (i.e., attention, learning, reaction time), experience and practice [23]. Impairment of the proprioceptive sensation can disrupt any of these six resources, which contributes to postural control (**Figure 2**). "*Sensory strategies*" are one of the most critical issues to discuss. Sensory information from somatosensory (tactile sense and proprioception), visual and vestibular systems must be integrated to interpret complex sensory environments for achieving postural control. Depending on the environmental conditions, the relative contribution of each sensory system changes, which is referred to as "sensory re-weighting" [24]. Healthy persons rely on somatosensory (70%), vision (10%), and vestibular (20%) information when standing on a stable surface in a well-lit environment (13). On the other hand, when standing on an unstable surface, due to decreased dependence on surface somatosensory inputs for postural orientation, they need to increase sensory weighting to vestibular and visual information [25]. The dynamic regulation or re-weighting of sensory cues is essential for maintaining postural stability when moving between different environments requiring distinct sensorial systems, such as different surfaces (i.e., walking on the sidewalk, walking on grass) or different lighting (i.e., moving in a well-lit room, moving in a dark room) [23]. The interplay between these three sensory modalities is critical for accurate estimates of self-motion and postural control [26].

Besides different sensory cues, different mechanical conditions provide significant advantages to humans for maintaining upright standing [27]. Decreased proprioception could lead to "*biomechanical constraints*" such as abnormal joint biomechanics and decreased muscle strength [28, 29], leading to postural dyscontrol. The "*control of dynamics*" is defined as the ability to perceive body segments relative to one another to stabilize the COM. Maintaining COM requires input from multiple sensory systems, sensorial re-weighting, and multisensory integration to calculate body state, including the COM and heading [30]. "*Movement strategies*" (i.e., postural

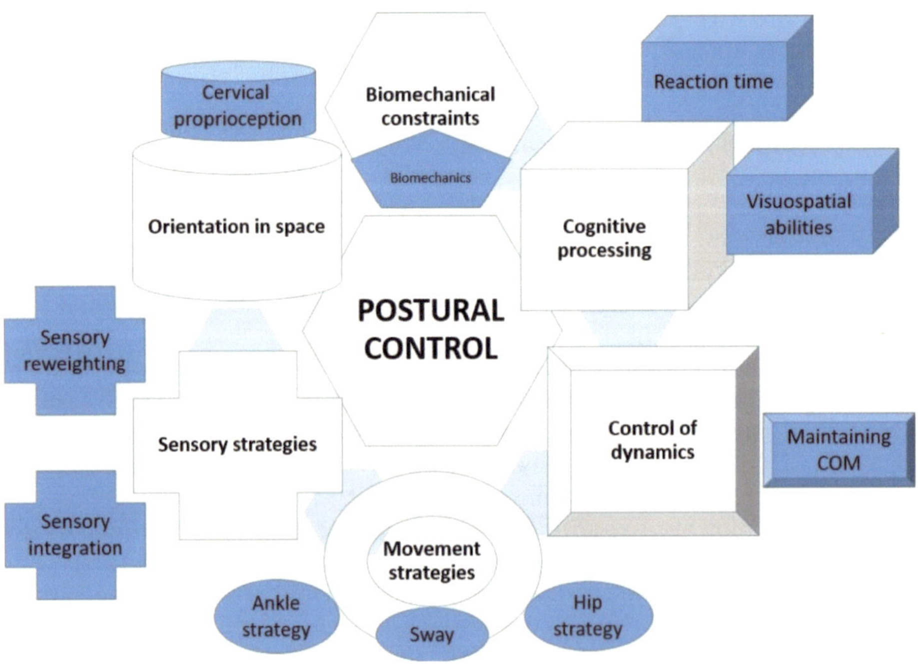

Figure 2.
Adapted from the framework of the six important resources required for postural control system by Horak, 2006 [23], the contribution of proprioception sensation to postural control.

sway, ankle strategy, hip strategy) can be used to return the body to equilibrium in a stance position [23]. Without proprioceptive input from the ankle and knee, ankle muscle responses are delayed suggesting that lower leg balance correcting responses are triggered by hip and, possibly, trunk proprioceptive inputs. Especially hip muscle proprioceptive inputs, considered critical for automatic balance correcting responses [31]. Additionally, cervical proprioception is of particular importance for "*orientation in space*". Neck muscle inflow has effects on the perception of body orientation and motion. Prolonged, intense proprioceptive input from neck muscles can induce persistent influences on self-motion perception and cognitive body representation [32]. The loss of proprioception could also impact the "*cognitive processing*" specifically the reaction time, and other factors such as attention, memory, and visuospatial abilities may contribute to spatial cognitive skills (**Figure 2**) [23].

2. Clinical implications and evaluation of proprioception

The loss of proprioceptive afferents may affect the control of muscle tone, disrupts postural reflexes, and severely impairs spatial and temporal aspects of movement [33]. Proprioceptive impairments are associated with various neurological conditions such as stroke [34], Parkinson's disease [35], peripheral neuropathy [36], as well as orthopedic conditions such as low back pain [37], neck pain [38], sports injuries like chronic ankle instability [39], ACL injuries [40], post-operatively such as mastectomy [41], knee arthroplasty [42], and aging [43]. Considering the importance of proprioception for motor control, a detailed evaluation of proprioceptive sense and application of treatment approaches focusing on training the proprioceptive sense is important for restoring motor function. Proprioception can be measured by using specific and non-specific tests in clinical practice.

Specific Tests of Proprioception: assess an individual's status regarding the joint position sense and kinesthesia [21]. *Joint position sense* tests assess precision or accuracy in repositioning the joint at a predetermined target angle and can be measured as active joint position detection (AJPD) [e.g., position matching task, position copying task] and passive joint position detection (PJPD) [e.g., thumb finding test, dual-joint position test] [44]. *Kinesthesia* tests assess the ability to perceive joint movement. For evaluating the perceptual aspect of proprioception, psychophysical thresholds represent the gold standard [33]. These tests are usually performed passively and can be measured by using passive motion detection threshold (PMDT) and passive motion direction discrimination (PMDD) [e.g., distal proprioception test, Rivermead Assessment of Somatosensory Perception] [44].

Non-specific Tests of Proprioception: for determining the contribution of proprioceptive signals on balance control, functional balance tests can be used to provide an estimate of potential proprioceptive disturbances [33]. These tests involve all body and other sensory and motor functions; therefore, they are considered non-specific tests of proprioception [21]. *Balance tests* can be modified to challenge proprioception, such as unilateral/bilateral stance with eyes open/closed, different supporting surfaces (i.e., stable or unstable), and with/without perturbations [44, 45]. *Stereognosis* and *skilled motor function tests* are important as they indicate the contribution of proprioceptive system in the performance of many activities of daily living [46].

3. Neurologic correlation

The complexity of sensorimotor systems requires deep knowledge of anatomy and physiology to analyze and localize the symptoms and the signs of the patients.

Joint sense and vibration sense examination is an important component of neurological examination.

The classic diseases causing sensory ataxia are tabes dorsalis, polyneuropathies (especially involving large fibers), dorsal root ganglionopathies and subacute combined degeneration. With parietal lobe lesion, position sense is often impaired and vibration preserved [5]. Vibratory sensation may also be impaired in lesions of the peripheral nerves, plexopathies, radiculopathies, dorsal root ganglion, posterior columns and medial lemniscus. In patients with peripheral neuropathies, vibration sensation is lost in the lower extremities first. Impaired vibration from posterior column disease is more likely to be uniform at all sites in the involved extremities. In spinal cord diseases, detecting a "level" of vibration sensory (segmental demarcation) loss over the spinous processes is crucial for diagnosis [5]. In patients with diabetic neuropathy, the decline in proprioceptive function may be caused by impairment in muscle spindle function and or the spindle receptors itself [47].

In patients with hereditary sensory and autonomic neuropathy type III patients (Riley-Day Syndrome, familial dysautonomia) ataxic gait is explained by poor proprioceptive acuity at the knee joint [48]. In mitochondrial ataxias sensory ataxia (which classically include gait ataxia worsened by loss of visual fixation) is due to the involvement of proprioception, secondary to peripheral neuropathy or neuronopathy [49]. In patients following whiplash type injuries involving soft tissues of cervical spine leads to proprioceptive deficits affecting head and position sense. Also in patients with chronic whiplash associated disorders are reported to have balance and dizziness problems, head and eye movement impairments reflecting mismatch od afferent input from the proprioceptive, visual and vestibular systems [8, 50]. Lesions of the dorsal columns impairs sensation of touch, vibration and proprioception in the ipsilateral side of the body below the injury level [51]. In patients with non-specific low back pain, postural control is impaired during standing and slow performance movements. This is due to an altered use of ankle compared to back proprioception related activity in right primary motor cortex and frontoparietal cortex [52]. Brainstem lesions resemble those in spinal cord disease as it selectively involves spinothalamic tract or medial lemniscus causing contralateral loss of position sense and vibration sense [5].

Neglect is a condition in which patients loose self-spatial awareness opposite to the damaged site of the brain. It is proposed that it is associated with the lesions of the dorsal stream causing dysfunction of proprioceptive space which is encoded in the bilateral parietal cortex [53]. Loss in the position sense may cause pseudochoreoathetosis as well. This abnormal involuntary, spontaneous movements are restricted to the parts with proprioceptive sensory loss. It is proposed that failure to integrate cortical proprioceptive sensory inputs in striatum may explain this situation [5, 54].

There are experimental evidence of proprioception impairments in Parkinson's disease. Parkinsonian gait is affected by the involvement of lower limb proprioceptive deficits as well as the involvement contralateral somatosensory and premotor lateral cortices and posterior cingulate cortex and basal ganglia and bilateral prefrontal cortex [10, 55, 56]. It was also shown that conscious perception of kinaesthetic stimuli is impaired in Parkinson's disease as cerebro-basal loops are not intact [9].

Weeks and colleagues showed that patients with cerebellar damage had reduced dynamic proprioceptive acuity which was also parallel to their motor deficits [3]. Diseases of the primary somatosensory cortex do not generally produce sensory symptoms but deteriorate fine and delicate manipulations in the contralateral part depending on position sense [2, 5]. Many patients with stroke experience proprioceptive deficits. Recovery of proprioception increases in the chronic phase [57, 58]. In study by Pope it was shown that proprioceptive input from the neck also may change cerebellar output affecting M1 plasticity [59]. In the study of Vidoni and colleagues preserved motor learning after stroke was related to the degree of proprioceptive deficit suggesting the relation between proprioceptive perception from muscle spindles and motor learning and central neuroplasticity [58, 60].

4. Proprioception after orthopedic surgeries

Studies on changes in joint proprioception after orthopedic surgeries are available in the literature. This section consists of the information in the literature about our five major joints.

4.1 Knee joint

Knee proprioception is necessary to achieve normal joint coordination during movement as well as providing joint stabilization [61, 62]. The anterior cruciate ligament (ACL), posterior cruciate ligament, collateral ligaments and menisci contribute to proprioception with the help of proprioceptors they have [63, 64]. The mechanoreceptors of the cruciate ligaments, together with the mechanoreceptors of the joint capsule, transmit information about the extension and flexion of the knee joint to the brain [65].

The ACL is the most important ligament involved in knee mechanical and neuromuscular stability. It contributes to proprioception in joint movement. However, the ACL is the most frequently injured ligament. After ACL rupture, knee proprioception is disrupted [66, 67].

Various autografts and allografts are used for ACL reconstruction. Patellar tendon or hamstring tendons may be preferred in patients using autografts. In addition, different techniques and materials are used. However, there is no gold standard in graft and technique selection [68]. In order for ACL reconstruction to be successful, not only mechanical but also neuromuscular stability is required. Neuromuscular stability depends on obtaining proprioception [69]. ACL injury leads to degradation of mechanoreceptors and a histologic study revealed that free nerve endings disappear after 1 year [70]. The effectiveness of ACL reconstruction in regaining proprioception has been tried to be revealed by some studies [71–74]. While some studies argue that ACL reconstruction is not sufficient to restore joint position [71–73], some studies advocate the adverse opinion [74]. The lack of a test to distinguish about whether the proprioception is derived from the soft tissues around the knee and capsule or from mechanoreceptors on ACL prevents to reach a certain decision about the mechanoreceptors of ACL [75].

Even after total knee arthroplasty, the contribution of the soft tissues around the knee to proprioception continues. In order to take advantage of this effect and ensure satisfactory outcomes in these patients, the soft tissue and gap must be well adjusted. Unicompartmental replacement protecting the ACL may be more advantageous in not reducing proprioception due to the proprioceptive effect of ACL. Also Ishii et al. [76] conclude that balance is improved after the postoperative period in bilateral total knee arthroplasty. It is stated that the first 6-week period is the critical period for adaptation time and proprioceptive loss after total knee replacement, and a new pattern in the knee load distribution occurs with postoperative rehabilitation [77].

4.2 Hip joint

Loss of proprioception, balance, sensation as joint position and kinesthetic are frequently observed in patients with knee osteoarthritis [78, 79]. Shakoor et al. [80] described significant sensory deficits associated with hip osteoarthritis, and these deficiencies involved both upper and lower limbs. The mechanism for this remains unclear; however, it has been suggested that there may be neurological feedback mechanisms or a inherent generalized neurological defect [78].

The greatest portion of mechanoreceptors and free nerve endings and highest concentration of pain receptors are located in the anterosuperior, posterosuperior and anterolateral labrum, respectively [81, 82].

There is no satisfactory information about proprioception impairment after surgeries due to hip pathologies. In the literature on the relationship between arthroplasty and proprioception, there are studies related to the knee rather than the hip. Interestingly, Ishii et al. [83] found no difference in proprioceptive responses among participants in the total hip arthroplasty, hemiarthroplasty and healthy control groups. They thought that the mechanoreceptors in the muscles, tendons and ligaments were responsible for joint proprioception rather than the intracapsular structures. While capsular receptors play a secondary role, muscle receptors play a primary role in hip proprioception. Therefore, it has been suggested that proprioception does not decrease after surgery, despite the capsule being removed during arthroplasty [84].

The effects of FAI and labral tear treatments on proprioception are not well known, but due to their proprioceptive properties, hip musculotendinous and capsuloligamentous tissues contribute to lower limb posture and stabilization through neuromuscular control. Therefore, preserving proprioceptive tissues as much as possible will prevent lower extremity injuries in arthroscopy operations.

4.3 Ankle joint

Ankle injuries are in the first place in sports-related injuries and lateral ankle sprains constitute the majority of this [85]. Unfortunately, many of these acute injuries can become chronic [86, 87]. Training, fatigue, and ankle injuries can affect ankle proprioception. Joint position sense, peroneal reaction time, EMG evaluation of peroneal muscles, and balance tests are tools to evaluate proprioception before and after ankle injuries or surgeries.

There are two important anatomical structures that provide proprioception and are located around the foot and ankle. Superior and inferior extensor retinaculum act as a pulley protecting tendons close to bony structures. The lateral ankle complex is the other anatomical structure with proprioceptive properties [88, 89]. Both acute and chronic injuries of the ankle can predispose the proprioceptors of the ankle. The differentiation in proprioception after these injuries were presented in the literature. While Vries et al. [90] stated that there was no difference between chronic ankle injury, acute trauma and healthy control groups, there are studies suggested that proprioception after acute inversion injuries and chronic ankle injuries are decreased [91–93]. Recovery of the proprioception is crucial after ankle injuries to maintain balance control. In order to achieve this, rehabilitation should not be neglected, especially after lateral ankle sprains.

A study conducted by Conti et al. [94] found no difference in proprioception between operated and non-operated side in total ankle arthroplasty. However, ankle arthroplasty has the worst outcome in terms of proprioception and balance compared to total hip and knee arthroplasty [95].

4.4 Shoulder joint

Some studies have revealed Pacinian corpuscles and Golgi tendon organ with mechanoreceptors in the shoulder [96, 97]. However, they discovered that while there are free nerve endings in the labrum and subacromial bursa, these structures do not contain mechanoreceptors. It is also thought that the supraspinatus muscle has more receptors than the infraspinatus muscle contains [98].

The pathological conditions of the shoulder joint can affect shoulder proprioception. Surgical shoulder diseases include rotator cuff tears, subcacromial pathologies, biceps tendon diseases and instabilities. Studies comparing pre- and post-surgical proprioception in the shoulder joint are not sufficient. In a study conducted by Aydın et al. [99], it was revealed that there was no difference in terms of proprioception between surgically treated and non-surgically treated shoulders in cases of instability. Duzgun et al. [100] stated a rapid recovery in shoulder joint proprioception after rotator cuff surgery as their experience.

Shoulder arthroplasty is thought to negatively affect proprioception. It has been stated that intervention to the subscapularis muscle and glenohumeral ligaments during shoulder arthroplasty may be effective in this decrease in proprioception [101, 102].

4.5 Elbow joint

Soft tissue damage is significant in elbow arthroplasty. Both flexor and extensor muscles are affected, collateral ligaments are released and capsule is removed. Therefore, the proprioceptive tissues as like skin, capsule, muscle and tendons are damaged. Despite the role of proprioception is still not well-established, one study was found an impairment in proprioception after total elbow arthroplasty [103].

In conclusion, proprioception may be adversely affected after joint surgeries. It should definitely be included in the rehabilitation program considering this situation. Proprioception seems to be an important factor for gaining balance and gait speed, especially after arthroplasties in the lower extremity.

Conflict of interest

The authors declare no conflict of interest.

Author details

Pinar Gelener[1]*, Gözde İyigün[2] and Ramadan Özmanevra[3]

1 Department of Neurology, Faculty of Medicine, University of Kyrenia, Kyrenia, Northern Cyprus, Mersin 10 via Turkey

2 Department of Physiotherapy and Rehabilitation, Faculty of Health Sciences, Eastern Mediterranean University, Famagusta, Northern Cyprus, Mersin 10 via Turkey

3 Department of Orthopedics and Traumatology, Faculty of Medicine, University of Kyrenia, Kyrenia, Northern Cyprus, Mersin 10 via Turkey

*Address all correspondence to: drpinargelener@gmail.com

References

[1] Hillier S, Immink M, Thewlis D. Assessing proprioception: a systematic review of possibilities. Neurorehabilitation and neural repair. 2015;29(10):933-949. DOI:10.1177/1545968315573055

[2] Gilman S. Joint position sense and vibration sense: anatomical organisation and assessment. Journal of Neurology, Neurosurgery & Psychiatry.2002;73:473-477.

[3] Weeks HM, Therrien AS, Bastian AJ. The cerebellum contributes to proprioception during motion. Journal of Neurophysiology.2017;118(2):693-702. DOI:10.1152/jn.00417.2016

[4] Cignetti F, Caudron S, Vaugoyeau M, Assaiante C. Body schema disturbance in adolescence: from proprioceptive integration to the perception of human movement. Journal of Motor Learning and Development.2013;1(3):49-58. DOI: 10.1123/jmld.1.3.49

[5] Campbell WW, DeJong RN. DeJong's the neurologic examination. Lippincott Williams & Wilkins; 2005.

[6] Bosco G, Poppele RE. Proprioception from a spinocerebellar perspective. Physiological reviews. 2001;81(2):539-568.

[7] MacKinnon CD. Sensorimotor anatomy of gait, balance, and falls. Handb Clin Neurol. 2018;159, 3-26. DOI: 10.1016/B978-0-444-63916-5.00001-X.

[8] Armstrong B, McNair P, Taylor D. Head and neck position sense. Sports medicine. 2008;38(2):101-117. DOI: 10.2165/00007256-200838020-00002

[9] Maschke M, Gomez CM, Tuite PJ, Konczak J. Dysfunction of the basal ganglia, but not the cerebellum, impairs kinaesthesia. Brain. 2003;126(10):2312-2322. DOI: 10.1093/brain/awg230

[10] Ribeiro L, Bizarro L, Oliveira A. Proprioceptive deficits in Parkinson's disease: from clinical data to animal experimentation. Psychology & Neuroscience. 2011;4(2):235-244. DOI:10.3922/j.psns.2011.2.009

[11] Luan H, Gdowski MJ, Newlands SD, Gdowski GT. Convergence of vestibular and neck proprioceptive sensory signals in the cerebellar interpositus. Journal of Neuroscience. 2013;16;33(3):1198-210. DOI: 10.1523/JNEUROSCI.3460-12.2013

[12] Brooks JX, Cullen KE. Multimodal integration in rostral fastigial nucleus provides an estimate of body movement. Journal of Neuroscience. 2009;29(34):10499-10511. DOI:10.1523/JNEUROSCI.1937-09.2009

[13] Lephart SM, Riemann BL, Fu FH. Introduction to the sensorimotor system. In: Lephart S M, Fu F H, Editors. Proprioception and Neuromuscular Control in Joint Stability. Human Kinetics; Champaign; 2000:37-51.

[14] Riemann BL, Lephart SM. The sensorimotor system, Part II: The role of proprioception in motor control and functional joint stability. J Athl Train. 2002;37(1):80-84. DOI:10.1016/j.jconhyd.2010.08.009

[15] Riemann BL, Lephart SM. The sensorimotor system, Part I: The physiologic basis of functional joint stability. J Athl Train. 2002;37(1):71-79.

[16] Valori I, McKenna-Plumley PE, Bayramova R, Callegher CZ, Altoè G, Farroni T. Proprioceptive accuracy in Immersive Virtual Reality: A developmental perspective. Senju A, ed. PLoS One. 2020;15(1):e0222253. DOI:10.1371/journal.pone.0222253

[17] Marshall AC, Gentsch A, Schütz-Bosbach S. The interaction between interoceptive and action states within a framework of predictive coding. Front Psychol. 2018;9(FEB):180. DOI:10.3389/fpsyg.2018.00180

[18] Martin TA, Keating JG, Goodkin HP, Bastian AJ, Thach WT. Throwing while looking through prisms II. Specificity and storage of multiple gaze-throw calibrations. Brain. 1996;119(Pt 4):1199-1211. DOI:10.1093/brain/119.4.1199

[19] Dakin CJ, Bolton DAE. Forecast or fall: Prediction's importance to postural control. Front Neurol. 2018;9:924. DOI:10.3389/fneur.2018.00924

[20] Maravita A, Spence C, Driver J. Multisensory integration and the body schema: Close to hand and within reach. Curr Biol. 2003;13(13):R531-R539. DOI:10.1016/S0960-9822(03)00449-4

[21] Röijezon U, Clark NC, Treleaven J. Proprioception in musculoskeletal rehabilitation: Part 1: Basic science and principles of assessment and clinical interventions. Man Ther. 2015;20(3):368-377. DOI:10.1016/j.math.2015.01.008

[22] Horak FB. Postural Control. In: Binder M.D., Hirokawa N., Windhorst U. (Eds). Encyclopedia of Neuroscience. Springer, Berlin, Heidelberg. ; 2009:3212-3219. DOI:10.1007/978-3-540-29678-2_4708

[23] Horak FB. Postural orientation and equilibrium: What do we need to know about neural control of balance to prevent falls? Age Ageing. 2006;35(SUPPL.2):7-11. DOI:10.1093/ageing/afl077

[24] Nashner L, Berthoz A. Visual contribution to rapid motor responses during postural control. Brain Res. 1978:403-407. DOI:10.1016/0006-8993(78)90291-3

[25] Peterka RJ. Sensorimotor integration in human postural control. J Neurophysiol. 2002;88(3):1097-1118. DOI:10.1152/jn.2002.88.3.1097

[26] Hwang S, Agada P, Kiemel T, Jeka JJ. Dynamic reweighting of three modalities for sensor fusion. PLoS One. 2014;9(1):e88132. DOI:10.1371/journal.pone.0088132

[27] Peterka RJ. Comparison of Human and Humanoid Robot Control of Upright Stance. J Physiol Paris. 2009;103(5):149-158. DOI:10.1016/j.jphysparis.2009.08.001

[28] Ito T, Sakai Y, Ito Y, Yamazaki K, Morita Y. Association Between Back Muscle Strength and Proprioception or Mechanoreceptor Control Strategy in Postural Balance in Elderly Adults with Lumbar Spondylosis. Healthcare. 2020;8(1):58. DOI:10.3390/healthcare8010058

[29] Ribeiro F, Oliveir J. Factors Influencing Proprioception: What do They Reveal? In: Biomechanics in Applications. InTech; 2011. DOI:10.5772/20335

[30] Chiba R, Takakusaki K, Ota J, Yozu A, Haga N. Human upright posture control models based on multisensory inputs; in fast and slow dynamics. Neurosci Res. 2016;104:96-104. DOI:10.1016/j.neures.2015.12.002

[31] Bloem B, Allum JHJ, Carpenter M, Verschuuren J, Honegger F. Triggering of balance corrections and compensatory strategies in a patient with total leg proprioceptive loss. Exp Brain Res. 2002;142(1):91-107. DOI:10.1007/s00221-001-0926-3

[32] Pettorossi VE, Schieppati M. Neck Proprioception Shapes Body Orientation and Perception of Motion. Front Hum Neurosci. 2014;8:1-13. DOI:10.3389/fnhum.2014.00895

[33] Aman JE, Elangovan N, Yeh IL, Konczak J. The effectiveness of proprioceptive training for improving motor function: A systematic review. Front Hum Neurosci. 2015;8(JAN):1-18. DOI:10.3389/fnhum.2014.01075

[34] Rand D. Proprioception deficits in chronic stroke—Upper extremity function and daily living. PLoS One. 2018;13(3):e0195043. DOI:10.1371/journal.pone.0195043

[35] Teasdale H, Preston E, Waddington G. Proprioception of the Ankle is Impaired in People with Parkinson's Disease. Mov Disord Clin Pract. 2017;4(4):524-528. DOI:10.1002/mdc3.12464

[36] Li L, Zhang S, Dobson J. The contribution of small and large sensory afferents to postural control in patients with peripheral neuropathy. J Sport Heal Sci. 2019;8(3):218-227. DOI:10.1016/j.jshs.2018.09.010

[37] Tong MH, Mousavi SJ, Kiers H, Ferreira P, Refshauge K, van Dieën J. Is There a Relationship Between Lumbar Proprioception and Low Back Pain? A Systematic Review With Meta-Analysis. Arch Phys Med Rehabil. 2017;98(1):120-136. DOI:10.1016/j.apmr.2016.05.016

[38] Stanton TR, Leake HB, Chalmers KJ, Moseley GL. Evidence of impaired proprioception in chronic, idiopathic neck pain: Systematic review and meta-analysis. Phys Ther. 2016;96:876-887. DOI:10.2522/ptj.20150241

[39] Xue X, Ma T, Li Q, Song Y, Hua Y. Chronic ankle instability is associated with proprioception deficits: A systematic review with meta-analysis. J Sport Heal Sci. 2020;00:1-10. DOI:10.1016/j.jshs.2020.09.014

[40] Kim HJ, Lee JH, Lee DH. Proprioception in Patients with Anterior Cruciate Ligament Tears: A Meta-analysis Comparing Injured and Uninjured Limbs. Am J Sports Med. 2017;45(12):2916-2922. DOI:10.1177/0363546516682231

[41] Zabit F, Iyigun G. A comparison of physical characteristics, functions and quality of life between breast cancer survivor women who had a mastectomy and healthy women. J Back Musculoskelet Rehabil. 2019;32(6):937-945. DOI:10.3233/BMR-181362

[42] Bragonzoni L, Rovini E, Barone G, Cavallo F, Zaffagnini S, Benedetti MG. How proprioception changes before and after total knee arthroplasty: A systematic review. Gait Posture. 2019;72:1-11. DOI:10.1016/j.gaitpost.2019.05.005

[43] Ribeiro F, Oliveira J. Aging effects on joint proprioception: The role of physical activity in proprioception preservation. Eur Rev Aging Phys Act. 2007;4:71-76. DOI:10.1007/s11556-007-0026-x

[44] Hillier S, Immink M, Thewlis D. Assessing Proprioception: A Systematic Review of Possibilities. Neurorehabil Neural Repair. 2015;29(10):933-949. DOI:10.1177/1545968315573055

[45] Clark NC, Ulrik R. Proprioception in musculoskeletal rehabilitation. Part 2 : Clinical assessment and intervention. Man Ther. 2015;20:378-387. DOI:10.1016/j.math.2015.01.009

[46] Stillman BC. Making sense of proprioception: The meaning of proprioception, kinaesthesia and related terms. Physiotherapy. 2002;88:667-676. DOI:10.1016/S0031-9406(05)60109-5

[47] Van Deursen RW, Sanchez MM, Ulbrecht JS, Cavanagh PR. The role of muscle spindles in ankle movement perception in human subjects with diabetic neuropathy. Experimental brain research. 1998 Apr 1;120(1):1-8.

[48] Bennell KL, Hinman RS, Metcalf BR, Crossley KM, Buchbinder R, Smith M, McColl G. Relationship of knee joint proprioception to pain and disability in individuals with knee osteoarthritis. Journal of orthopaedic research. 2003 Sep;21(5):792-797. DOI:10.1016/S0736-0266(03)00054-8

[49] Vernon HJ, Bindoff LA. Mitochondrial ataxias. InHandbook of clinical neurology 2018 Jan 1 (Vol. 155, pp. 129-141). Elsevier. DOI:10.1016/B978-0-444-64189-2.00009-3.

[50] Treleaven J, Peterson G, Ludvigsson ML, Kammerlind AS, Peolsson A. Balance, dizziness and proprioception in patients with chronic whiplash associated disorders complaining of dizziness: a prospective randomized study comparing three exercise programs. Manual therapy. 2016 Apr 1;22:122-130. DOI:10.1016/j.math.2015.10.017

[51] Wirz M, van Hedel HJ. Balance, gait, and falls in spinal cord injury. InHandbook of clinical neurology 2018 Jan 1 (Vol. 159, pp. 367-384). Elsevier. DOI:10.1016/B978-0-444-63916-5.00024-0.

[52] Goossens N, Janssens L, Caeyenberghs K, Albouy G, Brumagne S. Differences in brain processing of proprioception related to postural control in patients with recurrent non-specific low back pain and healthy controls. NeuroImage: Clinical. 2019 Jan 1;23:101881. DOI:10.1016/j.nicl.2019.101881

[53] Vakalopoulos C. Unilateral neglect: a theory of proprioceptive space of a stimulus as determined by the cerebellar component of motor efference copy (and is autism a special case of neglect). Medical hypotheses. 2007 Jan 1;68(3):574-600. DOI:10.1016/j.mehy.2006.08.013.

[54] Sharp FR, Rando TA, Greenberg SA, Brown L, Sagar SM. Pseudochoreoathetosis: movements associated with loss of proprioception. Archives of neurology. 1994 Nov 1;51(11):1103-1109. DOI:10.1001/archneur.1994.00540230041010

[55] Jacobs JV, Horak FB. Abnormal proprioceptive-motor integration contributes to hypometric postural responses of subjects with Parkinson's disease. Neuroscience. 2006 Jan 1;141(2):999-1009. DOI:10.1016/j.neuroscience.2006.04.014

[56] Keijsers NL, Admiraal MA, Cools AR, Bloem BR, Gielen CC. Differential progression of proprioceptive and visual information processing deficits in Parkinson's disease. European Journal of Neuroscience. 2005 Jan;21(1):239-248. DOI:10.1111/j.1460-9568.2004.03840.x

[57] Bowden JL, Lin GG, McNulty PA. The prevalence and magnitude of impaired cutaneous sensation across the hand in the chronic period post-stroke. PloS one. 2014 Aug 14;9(8):e104153. DOI:10.1371/journal.pone.0104153.

[58] Tashiro S, Mizuno K, Kawakami M, Takahashi O, Nakamura T, Suda M, Haruyama K, Otaka Y, Tsuji T, Liu M. Neuromuscular electrical stimulation-enhanced rehabilitation is associated with not only motor but also somatosensory cortical plasticity in chronic stroke patients: an interventional study. Therapeutic advances in chronic disease. DOI:10:2040622319889259.

[59] Popa T, Hubsch C, James P, Richard A, Russo M, Pradeep S, Krishan S, Roze E, Meunier S, Kishore A. Abnormal cerebellar processing of the neck proprioceptive information drives dysfunctions in cervical dystonia. Scientific reports. 2018 Feb 2;8(1):1-0. DOI:10.1038/s41598-018-20510-1

[60] Vidoni ED, Boyd LA. Preserved motor learning after stroke is related

to the degree of proprioceptive deficit. Behavioral and Brain Functions. 2009 Dec 1;5(1):36. DOI:10.1186/1744-9081-5-36.

[61] Pai YC, Rymer WZ, Chang RW, et al. Effect of age and osteoarthritis on knee proprioception. Arthritis Rheum. 1997;40:2260-2265. DOI: 10.1002/art.1780401223.

[62] Abelew TA, Miller MD, Cope TC, et al. Local loss of proprioception results in disruption of interjoint coordination during locomotion in the cat. J Neurophysiol. 2000;84:2704-2714. DOI: 10.1152/jn.2000.84.5.2709.

[63] Solomonow M, Krogsgaard M. Sensorimotor control of knee stability. A review. Scand J Med Sci Sports. 2001;11:64-80. DOI: 10.1034/j.1600-0838.2001.011002064.x.

[64] Jennings AG. A proprioceptive role for the anterior cruciate ligament: a review of the literature. J Orthop Rheumatol. 1994;7:3-13.

[65] Lephart SM, Riemann BL, Fu FH. Introduction to the sensorimotor system. In: Lephart SM, Fu FH, editors. Proprioception and neuromuscular control in joint stability. Champaign, IL: Human Kinetics;2000.

[66] Anders JO, Venbrocks RA, Weinberg M. Proprioceptive skills and functional outcome after anterior cruciate ligament reconstruction with a bone-tendon-bone graft. Int Orthop. 2008;32:627-633. DOI: 10.1007/s00264-007-0381-2

[67] Angoules AG, Mavrogenis AF, Dimitriou R, et al. Knee proprioception following ACL reconstruction: a prospective trial comparing hamstrings with bone-patellar tendon-bone autograft. Knee. 2011;18:76-82. DOI: 10.1016/j.knee.2010.01.009.

[68] Yosmaoglu HB, Baltacı G, Kaya D, et al. Comparison of functional outcomes of two anterior cruciate ligament reconstruction methods with hamstring tendon graft. Acta Orthop Traumatol Turc. 2011;45:240-247. DOI: 10.3944/AOTT.2011.2402.

[69] Krogsgaard MR, Dyhre-Poulsen P, Fischer-Rasmussen T. Cruciate ligament reflexes. J Electromyogr Kinesiol. 2002;12:177-182. DOI: 10.1016/s1050-6411(02)00018-4.

[70] Denti M, Monteleone M, Berardi A, et al. Anterior cruciate ligament mechanoreceptors. Histologic studies on lesions and reconstruction. Clin Orthop Relat Res. 1994;308:29-32. PMID: 7955696

[71] MacDonald PB, Hedden D, Pacin O, et al. Proprioception in anterior cruciate ligament-deficient and reconstructed knees. Am J Sports Med. 1996;24:774-8. DOI: 10.1177/036354659602400612

[72] Yosmaoglu HB, Baltacı G, Kaya D, et al. Tracking ability, motor coordination, and functional determinants after anterior cruciate ligament reconstruction. J Sport Rehabil. 2011;20:207-218. DOI: 10.1123/jsr.20.2.207

[73] Yosmaoglu HB, Baltacı G, Ozer H, et al. Effects of additional gracilis tendon harvest on muscle torque, motor coordination, and knee laxity in ACL reconstruction. Knee Surg Sports Traumatol Arthrosc. 2011;19:1287-1292. DOI: 10.1007/s00167-011-1412-5

[74] Reider B, Arcand MA, Diehl LH, et al. Proprioception of the knee before and after anterior cruciate ligament reconstruction. Arthroscopy. 2003;19:2-12. DOI: 10.1053/jars.2003.50006.

[75] Hogervorst T, Brand R. Mechanoreceptors in joint function. J Bone Joint Surg Am. 1998;80:1365-1378. DOI: 10.2106/00004623-199809000-00018.

[76] Ishii Y, Noguchi H, Takeda M, Sato J, Kishimoto Y, Toyabe S-I. Changes of body balance before and after total knee arthroplasty in patients who suffered from bilateral knee osteoarthritis. J Orthop Sci. 2013;18(5):727-732. DOI: 10.1007/s00776-013-0430-1.

[77] Thewlis D, Hillier S, Hobbs SJ, Richards J. Preoperative asymmetry in load distribution during quite stance persist following total knee arthroplasty. Knee Surg Sports Traumatol Arthrosc. 2014;22(3):609-614. DOI: 10.1007/s00167-013-2616-7.

[78] Hurley MV. The role of muscle weakness in the pathogenesis of osteoarthritis. Rheum Dis Clin N Am. 1999;25:283-298.

[79] Nyland J, Wera J, Henzman C, et al. Preserving knee function following osteoarthritis diagnosis: a sustainability theory and social ecology clinical commentary. Phys Ther Sport. 2015;16:3-9. DOI: 10.1016/j.ptsp.2014.07.003

[80] Shakoor N, Lee KJ, Fott LF, et al. Generalized vibratory deficits in osteoarthritis of the hip. Arth Rheum. 2008;59:1237-1240. DOI: 10.1002/art.24004

[81] Alzaharani A, Bali K, Gudena R, et al. The innervation of the human acetabular labrum and hip joint: an anatomic study. BMC Musculoskelet Disord. 2014;15:41. DOI:10.1186/1471-2474-15-41.

[82] [82]Haversath M, Hanke J, Landgraeber S, et al. The distribution of nociceptive innervation in the painful hip: a histological investigation. Bone Joint. 2013;J95:770-776. DOI: 10.1302/0301-620X.95B6.30262.

[83] Ishii Y, Tojo T, Terajima K, et al. Intracapsular components do not change hip proprioception. J Bone Joint Surg Br. 1999;81:345-348. DOI: 10.1302/0301-620x.81b2.9104.

[84] Nallegowda M, Singh U, Bhan S, Wadhwa S, Handa G, Dwivedi SN. Balance and gait in total hip replacement: a pilot study. Am J Phys Med Rehabil. 2003;82(9):669-677. DOI: 10.1097/01.PHM.0000083664.30871.C8.

[85] Janssen KW, Kamper SJ. Ankle taping and bracing for proprioception. Br J Sports Med. 2013;47:527-528. DOI: 10.1136/bjsports-2012-091836

[86] Karlsson J, Lansinger O. Lateral instability of the ankle joint. Clin Orthop Relat Res. 1992;276:253-261. PMID: 1537162

[87] Kerkhoffs GM, Handoll HH, de Bie R, Rowe BH, Struijs PA. Surgical versus conservative treatment for acute injuries of the lateral ligament complex of the ankle in adults. Cochrane Database Syst Rev. 2007;(2):CD000380. DOI: 10.1002/14651858.CD000380.pub2.

[88] Li HY, Zheng JJ, Zhang J, Cai YH, Hua YH, Chen SY. The improvement of postural control in patients with mechanical ankle instability after lateral ankleligaments reconstruction. Knee Surg Sports Traumatol Arthrosc. 2016;24(4):1081-1085. DOI: 10.1007/s00167-015-3660-2.

[89] Hertel J. Sensorimotor deficits with ankle sprains and chronic ankle instability. Clin Sports Med. 2008;27:353-70. DOI: 10.1016/j.csm.2008.03.006.

[90] Vries S, Kingma I, Blankevoort L, van Dijk CN. Difference in balance measures between patients with chronic ankle instability and patients after an acute ankle inversion trauma. Knee Surg Sports Traumatol Arthrosc. 2010;18(5):601-606. DOI: 10.1007/s00167-010-1097-1.

[91] Hertel J. Functional anatomy, pathomechanics, and pathophysiology of lateral ankle instability. J Athl Train. 2002;37:364-375. PMID: 12937557

[92] Docherty CL, Valovich McLeod TC, Shultz SJ.Postural control deficits in participants with functional ankle instability as measured by the balance error scoring system. Clin J Sport Med. 2006;16:203-208. DOI: 10.1097/00042752-200605000-00003.

[93] Jerosch J, Hoffstetter I, Bork H, Bischof M. The influence of orthoses on the proprioception of the ankle joint. Knee Surg Sports Traumatol Arthrosc. 1995;3:39-46. DOI: 10.1007/BF01553524.

[94] Conti SF, Dazen D, Stewart G, Green A, Martin R, Kuxhaus L, et al. Proprioception after total ankle arthroplasty. Foot Ankle Int. 2008;29(11):1069-1073. DOI: 10.3113/FAI.2008.1069.

[95] Butler RJ, Thiele RAR, Barnes CL, Bolognesi MP, Queen RM. Unipedal balance is affected by lower extremity joint arthroplasty procedure 1 year following surgery. J Arthroplasty. 2015;30(2):286-289. DOI: 10.1016/j.arth.2014.08.031.

[96] Vangness CT Jr, Ennis M, Taylor JG, Atkinson R. Neural anatomy of the glenohumeral ligaments, labrum, and subacromial bursa. Arthroscopy. 1995;11(2):180-184. DOI: 10.1016/0749-8063(95)90064-0.

[97] Ide K, Shirai Y, Ito H, Ito H. Sensory nerve supply in the human subacromial bursa. J Shoulder Elb Surg. 1996;5:371-382. DOI: 10.1016/s1058-2746(96)80069-3.

[98] Windhorst U. Muscle proprioceptive feedback and spinal networks. Brain Res Bull. 2007;73:155-202. DOI: 10.1016/j.brainresbull.2007.03.010.

[99] Aydin T, Yildiz Y, Yanmiş I, Yildiz C, Kalyon TA. Shoulder proprioception: a comparison between shoulder joint in healthy and surgically repaired shoulders. Arch Orthop Trauma Surg. 2001;121(7):422-425. DOI: 10.1007/s004020000245.

[100] Duzgun, I., & Turhan, E. (2017). Proprioception After Shoulder Injury, Surgery, and Rehabilitation. Proprioception in Orthopaedics, Sports Medicine and Rehabilitation, 35-45. DOI:10.1007/978-3-319-66640-2_4.

[101] Maier MW, Niklasch M, Dreher T, Wolf SI, Zeifang F, Loew M, Kasten P. Proprioception 3 years after shoulder arthroplasty in 3D motion analysis: a prospective study. Arch Orthop Trauma Surg. 2012;132(7):1003-1010. DOI: 10.1007/s00402-012-1495-6

[102] Kasten P, Maier M, Retting O, Raiss P, Wolf S, Loew M. Proprioception in total, hemi- and reverse shoulder arthroplasty in 3D motion analyses: a prospective study. Int Orthop. 2009;33(6):1641-1647. DOI: 10.1007/s00264-008-0666-0.

[103] Lubiatowski P, Olczak I, Lisiewicz E, Ogrodowicz P, Bręborowicz M, Romanowski L. Elbow joint position sense after total elbow arthroplasty. J Shoulder Elbow Surg. 2014;23(5):693-700. DOI: 10.1016/j.jse.2014.01.016.

Chapter 3

Recording of Proprioceptive Muscle Reflexes in the Lower Extremity

Juhani Partanen, Urho Sompa and Miguel Muñoz-Ruiz

Abstract

Electromyography (EMG) is routinely used in diagnostics of root syndromes in the lower extremity. By studying signs of axonal damage of different root levels in the corresponding myotomes of the lower extremity and back muscles with needle EMG reveals, which of the motor roots are injured in patients with suspected root compression. But by EMG study only injuries of the anterior motor roots are diagnosed. Routine electroneuromyography does not disclose specific injury of the afferent sensory posterior roots. However, the integrity of some the posterior roots is readily studied with myotatic reflexes. We have routinely measured a proprioceptive reflex, the H-reflex of the soleus muscle with stimulation of the posterior tibial nerve, and found it to be useful in the diagnostics of the S1 root syndrome. It seems to be possible to record H-reflex of the peroneus longus muscle at the L5 level. We discuss the serious problems with volume conduction, when trials to measure proprioceptive reflexes of the L4 and L5 levels are performed. It may also be useful to record the medium latency reflexes in the area of the posterior tibial nerve, which seems to have a different reflex arch (II-afferents – β-efferents) from H-reflex (Ia afferents – α efferents). These measurements are non-invasive and not time consuming, and we hope to be able to add them for the routine ENMG diagnostics, when appropriate.

Keywords: proprioception, Ia afferent, II afferent, alpha motor neuron, beta motor neuron, electromyography, H-reflex, muscle reflex, root compression

1. Introduction

Root syndrome diagnostics of the lower extremity is based on the clinical picture, anamnesis, symptoms and signs of the disease. Diagnostic investigations should be considered in a few weeks if symptoms are not resolved, or even earlier if paraparesis or bowel or bladder symptoms develop. Current imaging studies are excellent, but there is the problem with non-symptomatic degenerative changes vs. relevant findings with respect to the acute symptoms, especially in middle-aged and old patients. ENMG has another problem: after acute onset of the disease the proprioceptive tendon- and H-reflexes change in a few days. However, the development of pathological spontaneous activity, fibrillations and positive sharp waves, indicating axonal injury in electromyography (EMG) may take 2-3 weeks and even longer in distal muscles of the leg and foot [1]. Loss of motor units during maximal voluntary contraction may be observed soon, but more distinct changes, such as increase of duration, complexity (polyphasic and jittering waveform), and amplitude of motor unit potentials after parallel reinnervation may take several weeks and even months to develop [2]. That is why ENMG studies are usually not performed until several weeks after the acute onset stage. The aim of this chapter in to describe some proprioceptive reflexes, which may be used in acute stage of the disease, when clear needle EMG findings are not yet discernible, and proprioceptive reflex measurements which may be further developed for the ENMG diagnostics of root syndromes of the lower extremity.

1.1 Routine electroneurography in root syndromes of the lower extremity

Electroneuromyography of root syndromes of the lower extremity tends to concentrate on function of motor nerve fibres. Needle EMG observes axonal damage with fibrillation potentials, and loss and sprouting alterations of motor unit potentials. Signs of axonal damage may be searched in different myotomes of the lower extremity and paraspinal muscles [2]. F-responses and amplitudes of the motor responses may give supplementary information. The sensory responses are not affected, if the root lesion is proximal to the sensory paraspinal ganglion.

1.2 Pain in root syndromes and the methods to study the posterior roots

However, pain is usually more prominent symptom in root syndromes than motor weakness. Pain may express itself in the dermatomes of different root levels but often pain symptoms are obscure. The pain pathway uses the posterior roots, which may have a separate or more prominent injury than the anterior motor roots. ENMG study involving only motor nerve fibres may not be sufficient for the proper diagnosis of a root syndrome. Methods for studying the integrity of the posterior roots are needed. Posterior root compression may cause activation of pain C-fibres, but this may not invariably change proprioceptive reflexes using sensory afferent pathways with myelinated nerve fibres. Dermatomal evoked responses have been used, but they have not got any wide popularity. The method is awkward and time-consuming and the cerebral responses are small. This method is not recommended for clinical use [3].

1.3 H-reflexes of the distal muscles

Proprioceptive reflexes, especially H-reflexes, which use the posterior root pathway are too seldom used in ENMG diagnostics [4]. The only reflex we have routinely measured in patients with root syndrome is H-reflex of the soleus muscle. Its recording is easy, non-invasive and rapid, and very useful in S1 root syndrome diagnostics. It may also be used as a part of measurements to study polyneuropathy, an entity that should also be evaluated when root syndromes are investigated. Damage of the S1 posterior root often abolishes the H reflex response or causes slight prolongation or diminution of the reflex response [4]. Compression and injury of the anterior root is observed as a marked prolongation of the latency and diminution of the response amplitude (**Figure 1**). The clinical use of the soleus H-reflex requires a comprehensive normal material, which comprises corrections for height and age and sex of the patient (**Table 1**) [5].

We have not used systemically any H-reflexes of L5 and L4 levels. The H-reflex of the anterior tibial muscle may be recorded with slight tonic voluntary contraction

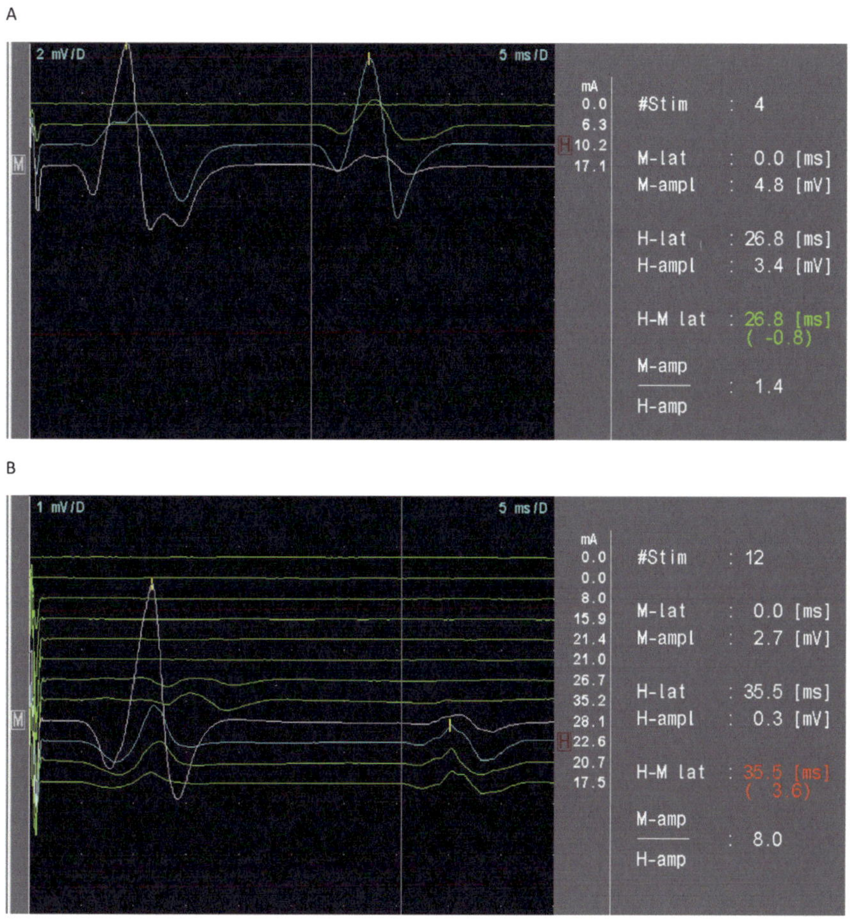

Figure 1.
*H-reflex in the healthy (A) and symptomatic (B) side of a patient with S1 root syndrome. Note the diminution of amplitude and large latency value of the H-reflex response in the symptomatic side, compared to the healthy one. The relatively large reflex asymmetry and good persistence also in the symptomatic side is consistent with S1 anterior root compression. The" H-M lat" describes the deviation in Z score value (normal <2) of the measured H-reflex latency from normal control values with height, age and sex corrections (**Table 1**). Note also the large difference in the M-amp/H-amp relationship.*

Distal latency (ms)	**SD**	**H-latency (ms)**		**SD**	
4.13	0.50	29,18		2.18	
Calculation of the presumed normativity and the expectation percentage: $(R^2) \times 100$ of the tibial H-reflex.					
$(R^2) \times 100$	SD	Constant	Height coefficient (hc)	Age coefficient (ac)	sex
48	1.562	2.110	0.160		
64	1.325	−6.239	0.193	0.085	
68	1.256	−15.210	0.247	0.094	−1.45
Calculation of the expectation value: constant + hc x height (cm) + ac x age + sex (male).					

Table 1.
Normal values and presumed normativity of the tibial H-reflex measured from the soleus muscle (surface electrodes near the border of the gastrocnemius muscle and the reference 2–3 cm distally).

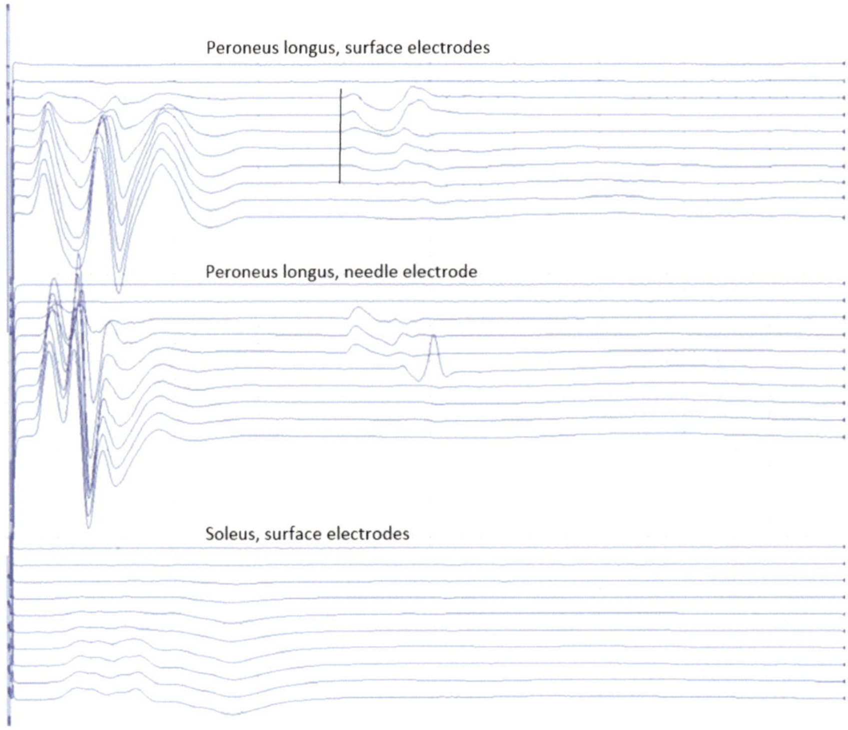

Figure 2.
H-reflex of peroneus longus, 32 years old male. Stimulation with 2 cm bipolar surface electrode to common peroneal nerve at the fibular head causing clear ankle dorsiflexion; ten concurrent stimulations with increasing stimulation current. Simultaneous recording of peroneus longus with both surface electrodes (interelectrode distance ca 3 cm), a 30 G concentric needle electrode, and soleus with surface electrodes (interelectrode distance ca 3 cm). Note the typical appearance of H-wave, latency 32 ms (vertical line), reaching its maximal amplitude before M-wave (contrary to the performance of F-responses) and appearance of H-wave solely on peroneus longus and not on soleus. 2 mV/div, 8 ms/div. A similar recording with surface electrodes on the lateral gastrocnemius muscle was also performed and no reflex response of this muscle was observed (not shown).

of the given muscle [4], but we have found it too difficult for routine use.
No H-reflex for the L5-level was described for clinical use in root syndromes. We have tried to measure H-reflexes of the peroneus longus and extensor hallucis longus muscles, but these measurements were hampered by volume conduction of reflexes of the triceps surae muscle. However, H-reflex of the peroneus longus muscle can be confirmed by recording it with EMG needle electrode (**Figure 2**). The peroneus longus H-reflex may disappear in the symptomatic side of a patient with unilateral L5 root syndrome (**Figure 3**). Problems with volume conduction are discussed at the end of this chapter.

1.4 Tendon reflexes and H-reflexes of the proximal muscles

The L3-4 posterior roots can be studied with the patellar reflex (**Figure 4a** and **b**), and the adductor reflex [6]. The adductor tendon reflex can be evoked by ipsilateral tap to the medial epicondyle of the femur. Surprisingly, this reflex may also be elicited easily by tap to several sites of the lower extremity: for instance contralateral patellar tap, as well as ipsi- and contralateral anterior superior iliac spine tap. On the contrary, patellar tendon reflex was obtained only by ipsilateral tap to the patellar tendon. H-reflex of the adductor muscle (latency 16.4 ms, SD 1.6) is obtained by percutaneous

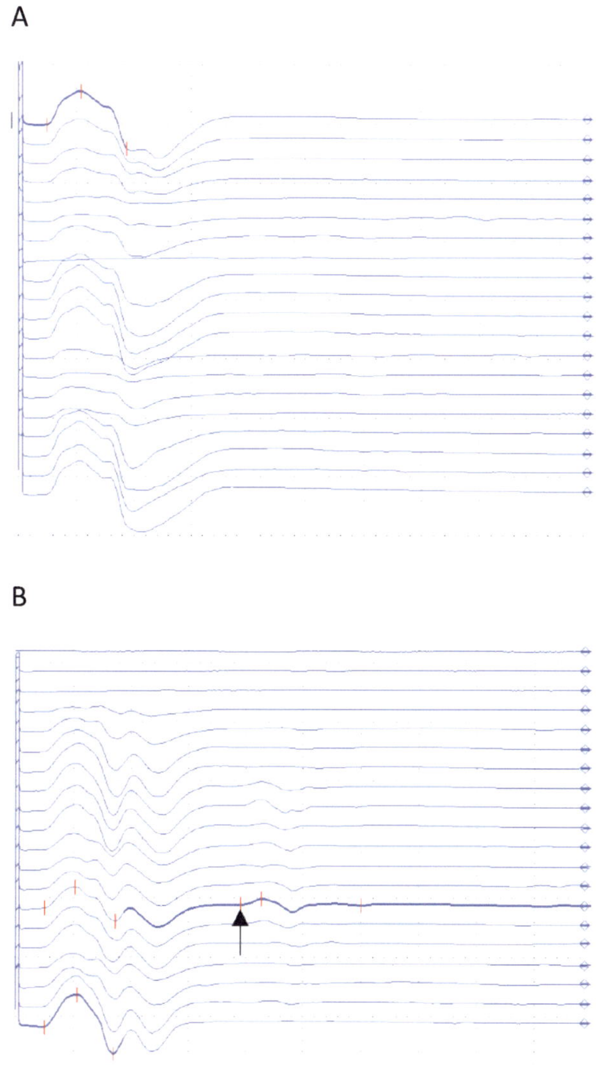

Figure 3.
A. M-response of the right peroneus longus muscle in a patient aged 80 with a right L5 root syndrome: Right L5 compression in MRI, positive needle EMG finding in the right L5 paraspinal muscle, F-response latency asymmetrically prolonged in the right extensor digitorum brevis muscle but bilaterally normal in the abductor hallucis muscles. Stimulation of the common peroneal nerve at the fibular head. Latency of the M-response 3.9 ms. The H-reflex cannot be elicited in spite of changes of the stimulation intensity 3.0-11.7 mA (submaximal and supramaximal). Calibration: 8 ms/div and 5 mV/div. No medium latency reflexes (see ***Figure 9****). The tibialis posterior nerve was evidently not coactivated with the stimulation of the common peroneal nerve, there was only plantar dorsiflexion. B. M-response and H-reflex (arrow) of the left peroneus longus muscle of the patient. Latency of the M-response 4.0 ms and H-reflex 31.2 ms. stimulation intensity 3.1-14.4 mA. Calibration as above.*

stimulation of the obturator nerve at the level of pubic tubercle. Even a medium-latency reflex, "late polysynaptic reflex response" of more than 50 ms was described in the adductor muscle [6]. However, these methods are rarely used in routine ENMG studies. The H-reflex of the quadriceps femoris muscle is readily recorded by stimulation of the femoral nerve [4], but we have not gained any experience with this method. The Achilles tendon reflex may also be recorded with surface electrodes on the triceps surae muscle. This recording was not used in routine ENMG studies. We should also remember that the tendon reflex and H-reflex have distinct differences [7].

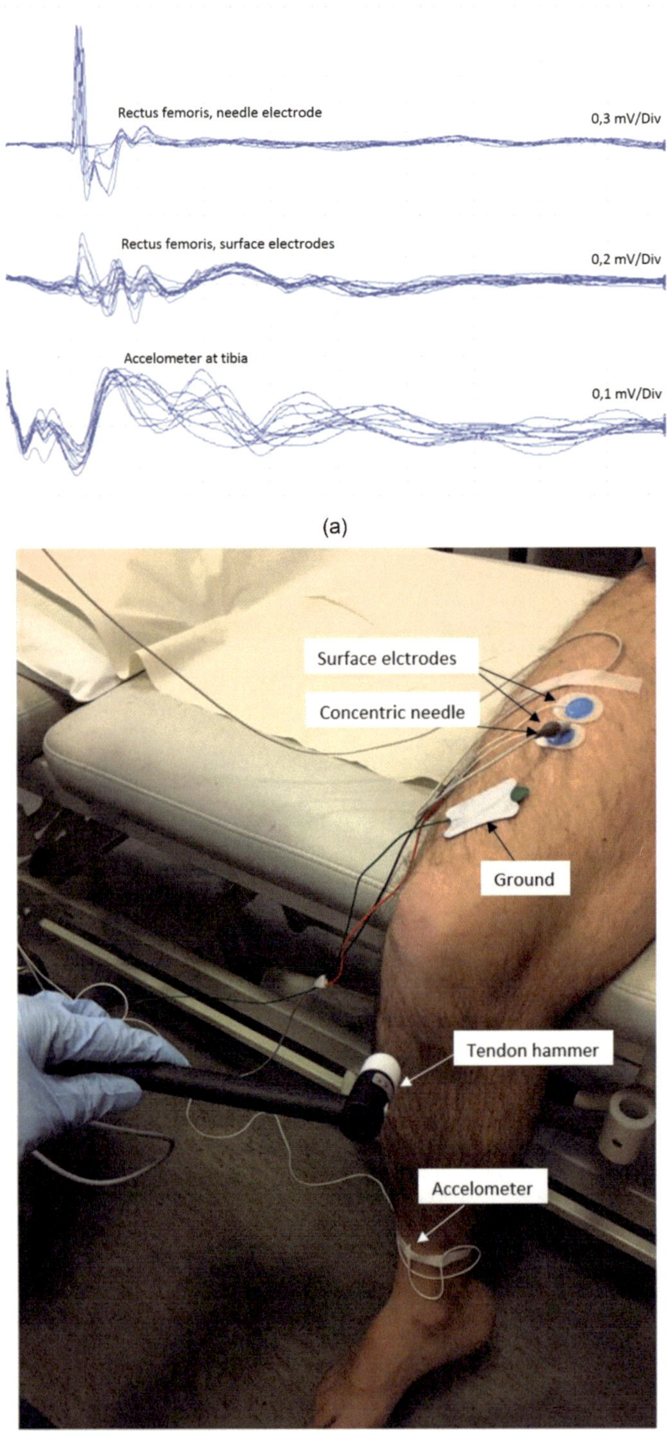

Figure 4.
a. Electrical recording of the patellar reflex, 33 years old male. Stimulation with tendon hammer electrically connected to EMG-machine. Recording is triggered by a strike to prepatellar tendon. Recording in the rectus femoris muscle with both concentric needle electrode (30 G) and surface electrodes and an accelerometer connected to tibia. Three separate recordings of same stimulation protocol. 7-10 superimposed responses. Patellar reflex at ca 20 ms. 20 ms/div. b. Electrical recording of the patellar reflex, responses shown in a.

1.5 Medium latency reflex responses

A medium latency reflex response (60-80 ms) of the soleus muscle can be recorded by supramaximal stimulus of the common peroneal nerve, which causes powerful twitch contraction of the peroneal muscles [8] (**Figure 5a**). Originally this reflex response was considered to use low-threshold muscle afferents and a transcranial loop, possibly involving the primary motor cortex and the supplementary motor area [9]. Later on it was demonstrated that the medium-latency reflex response of the soleus muscle to stretch does not involve a long reflex loop [10]. Soleus stretch resulting from unexpected perturbation during human walking elicits both short and medium latency reflex responses. It was concluded by cooling, ischaemia and tizanidine studies that the afferent receptors of the short latency component are Ia afferents and those of the medium latency component are II-afferents, respectively [11].

By stimulation of the common peroneal nerve at the fibular neck, only the medium latency reflex response can be recorded in electroneurography of the human soleus muscle [8] (**Figure 5a**). It was observed that stimulation of the common peroneal nerve results in long lasting (up to 200 ms) soleus H-reflex depression [12]. On the contrary, by stimulation of the posterior tibial nerve at the popliteal space, no medium latency reflex response can be recorded from the anterior tibial muscle [8].

1.6 Calculations of conduction velocities and the role of β-efferents in the medium latency reflexes

The distances between the stimulation and recording sites were measured when the responses depicted in **Figure 5a** were recorded. The afferent pathway for H-reflex latency 27.0 ms was 640 mm between the stimulation site at the popliteal space, and L1 spinal level. Respectively, the distance of the efferent pathway between L1 spinal level and the estimated motor point of the soleus muscle was 750 mm. Considering that the synaptic delay in the spinal cord is about 1 ms [7] we can conjecture that the afferent conduction time from the stimulation site to the spinal cord is 11 ms and the efferent conduction time is 15 ms. The respective conduction velocities are for Ia afferents 58 m/s and for α motor efferents 50 m/s. These values match well with the recordings of Ia afferent conduction velocity 64 m/s and α motor conduction velocity 56 m/s of the median nerve [13], assuming that the respective values are slightly slower in the lower than in the upper extremity. The more distally recorded Ia afferent conduction velocity between the popliteal fossa and ankle is 56 m/s [7]. Cutaneous afferents are slower than Ia afferents, 61 m/s in the upper extremity [13] and 48 m/s in the lower extremity [7].

A similar calculation may be performed for the medium latency reflex latency 62 ms. The distance from the proximal part of the soleus muscle (site of the most proximal muscle spindles) to the L1 spinal level was 670 mm, and the distance from L1 to the motor point of the soleus muscle was 750 mm. The estimated afferent conduction time is 30 ms and the efferent conduction time 31 ms, the spinal synaptic delay time was again estimated to be 1 ms. By these values we may calculate, that the afferent conduction velocity for II-afferent pathway is 22 m/s, and for the efferent conduction velocity is 24 m/s. This afferent conduction velocity matches well with the II-afferent conduction velocity 21 m/s observed in the lower extremity [14]. But the efferent conduction velocity 24 m/s is far too slow for the α motor efferent pathway, which was calculated to be 50 m/s in the H-reflex arch (see above).

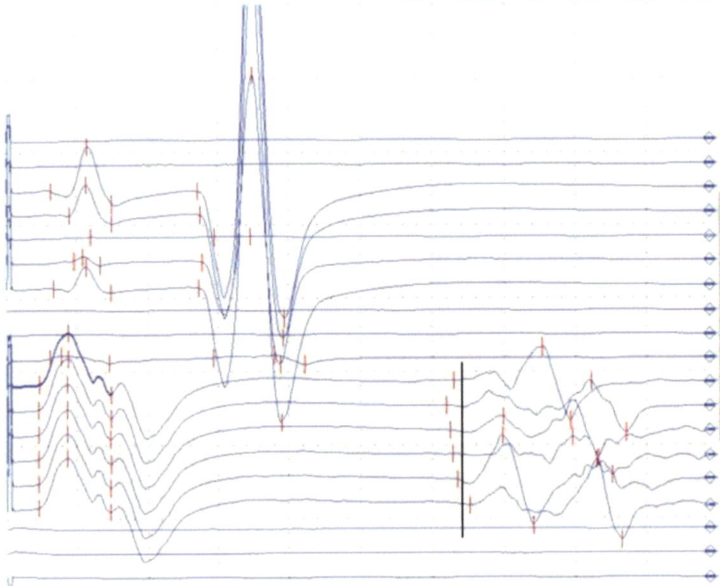

(a)

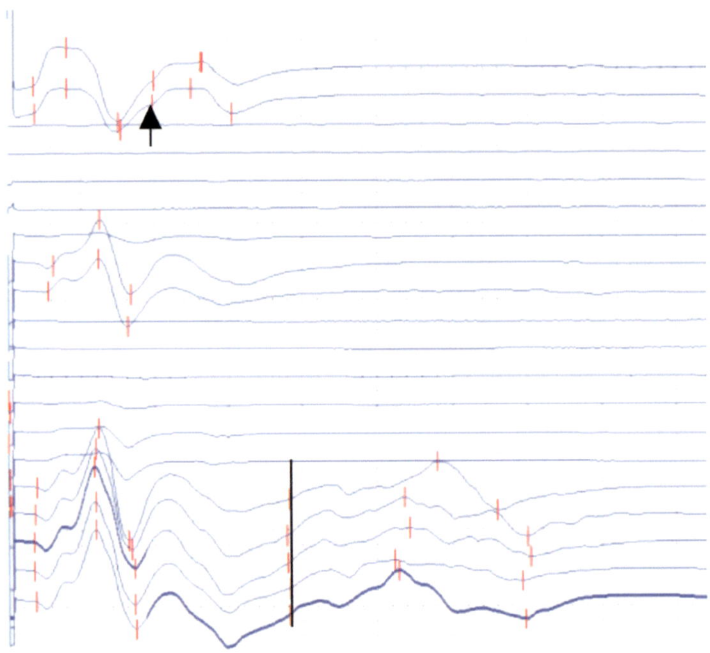

(b)

Figure 5.
a. the tibial H-reflex of the soleus muscle elicited with submaximal stimuli, minimum latency 27.0 ms. The stimulation was changed to the common peroneal nerve at the knee joint and supramaximal stimuli elicited the medium latency reflexes of soleus, minimum latency 62.0 ms (vertical line). The" M-response" was reflected from the pretibial muscles. For calculations of the afferent and efferent conduction velocities of the reflex responses see text. *Calibration: 10 ms/div, 2 mV/div. A voluntary healthy subject, male, age 31 y, height 166 cm. b. H-reflex of the median nerve (2 uppermost sweeps), latency 15,5 ms (arrow), stimulation: Median nerve at the elbow, recording with surface electrodes on the forearm flexors. The stimulation was changed to the radial nerve at the spiral groove (2 middle responses). When the stimulus was turned to supramaximal value, medium latency reflex responses, latency 30.5 ms, were elicited (vertical line, 5 lowermost sweeps) from the forearm flexors. Calibration: 8 ms/div, 2 mV/div. A voluntary healthy subject, male, age 31 y, height 166 cm.*

This fact justifies the hypothesis that the efferent pathway of the medium latency reflexes consists of skeletofusimotor β motor fibres, which are thinner and slower than α motor fibres. β motor efferents have been observed in man [15].

1.7 The possible influence of inhibitory pathways on the reflex responses

Ib afferent nerve fibres from Golgi tendon organs are slightly smaller than those of Ia afferents [11]. The electrically evoked excitatory postsynaptic potential may be curtailed by the inhibitory postsynaptic potential of only slightly longer latency than the excitatory postsynaptic potential [7]. There is a Ib inhibitory volley from the Golgi tendon organs, which originate from the proximal tendon insertion of the anterior tibial muscle, elicited by the strong contraction of the muscle by stimulation of the posterior tibial nerve at the popliteal space. This inhibitory volley may reach the spinal cord and prevent the occurrence of the medium latency reflex response. The lack of medium latency reflex was pointed out in this muscle [8]. Unexpected perturbation during walking elicits short- and medium-latency soleus reflex responses [11]. However, soleus stretch, caused by electric stimulation of the common peroneal nerve and powerful contraction of the pretibial muscles, elicits only a medium-latency reflex response of the soleus muscle. It may be considered that the Ia reciprocal inhibitory influence [7] plays a role in inhibition of the soleus short-latency reflex response in this situation.

1.8 Comparison with the upper extremity

The forearm flexor muscles (for example m. flexor carpi radialis and m. flexor digitorum superficialis) show H-reflexes, when the median nerve is stimulated at the elbow [4]. When the stimulation is changed to the radial nerve at the spiral groove, a medium reflex response may be recorded at the same site than the median H-reflex (**Figure 5B**). Thus, the respective reflex responses seem to be elicited in

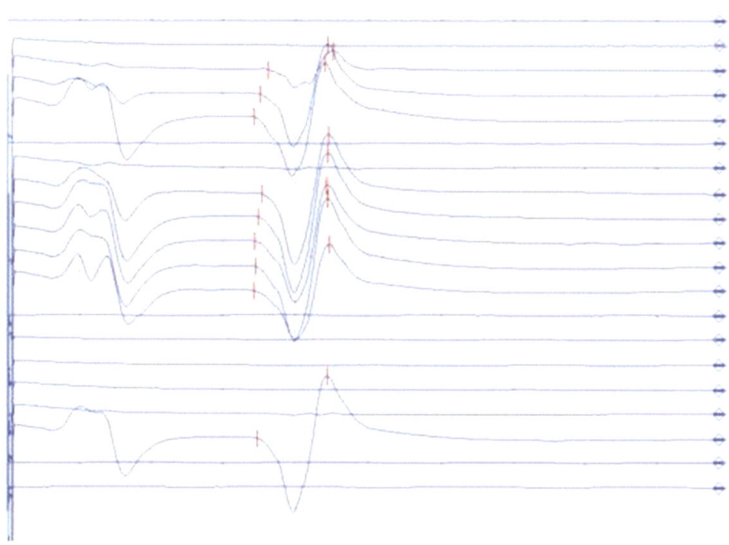

Figure 6.
*The" reflex" response of the anterior tibial muscle (latency 27.4 ms), recorded by the stimulation of the posterior tibial nerve at the popliteal fossa. Superficially it may be reminiscent to a myotatic reflex of the anterior tibial muscle, but in reality it is the H-reflex of the triceps surae muscle, volume conducted to the recording site (compare with the H-reflex recording in **Figure 4**). The" M-response" points out the direct activation of the triceps surae muscle. Calibration: 8 ms/div, 2 mV/div.*

A

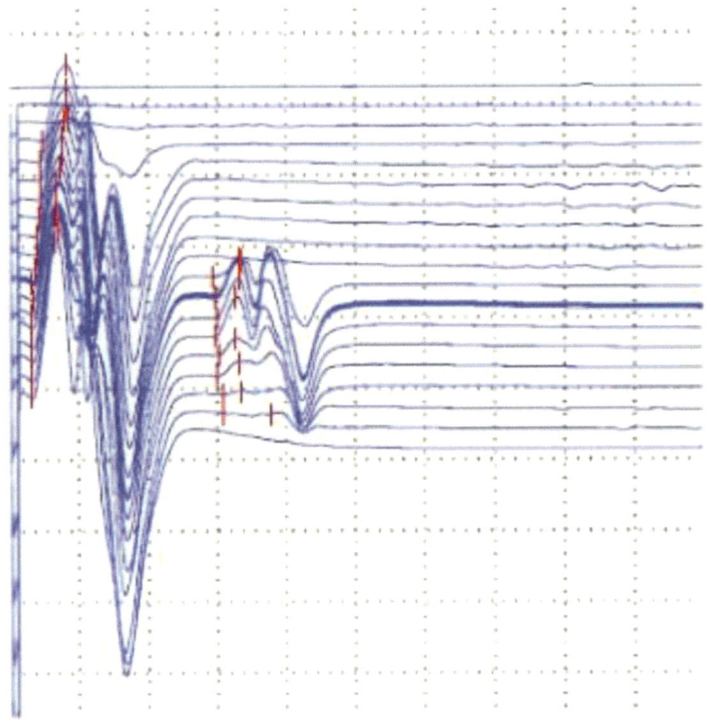

B

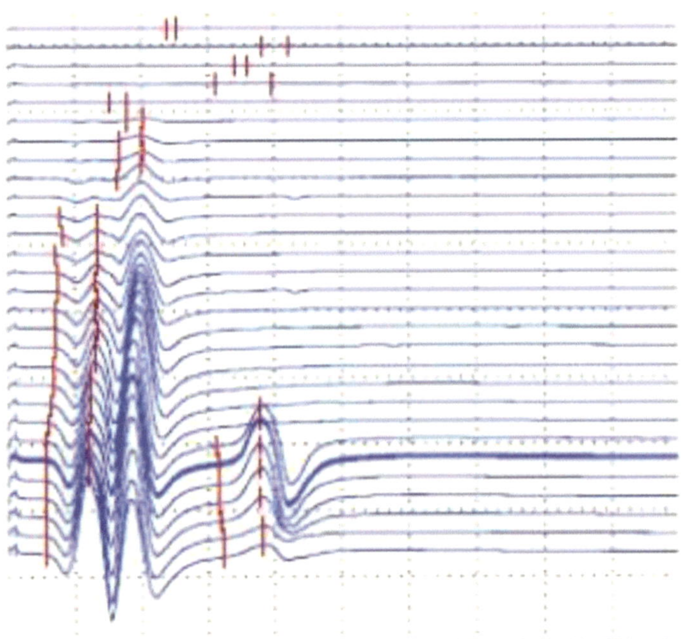

Figure 7.
"H-reflexes" of the A) peroneus longus and B) extensor hallucis longus muscles by stimulation of the common peroneal nerve and recorded with surface electrodes. In reality" H-reflexes" may be H-reflexes of the triceps surae muscle caused by spreading of stimuli to a branch of the posterior tibial nerve. Calibration 10 ms/div, 3 mV/div.

the upper extremity than in the lower one as well. As in the lower extremities the proprioceptive reflexes are important in the process of walking and running, they might be related to grip and climbing functions in the upper extremities and thus may have served an important role in primate evolution.

1.9 Problems with volume conduction

When the common peroneal nerve is stimulated, the stimulus spreads readily to the motor branches of the posterior tibial nerve. Thus, volume conduction is a source of error especially when the reflexes of the pretibial muscles are recorded with surface electrodes. The stimulation spreading to branches of the posterior tibial nerve may elicit H-reflex of the triceps surae muscle, recorded with electrodes on the surface of the anterior tibial muscle. This reflex response may imitate the myotatic reflex of the anterior tibial muscle (**Figure 6**). A similar problem may be encountered by recording of responses of the peroneus longus and extensor hallucis longus muscles (**Figure 7**), as well as the foot muscles (**Figure 8**). However a

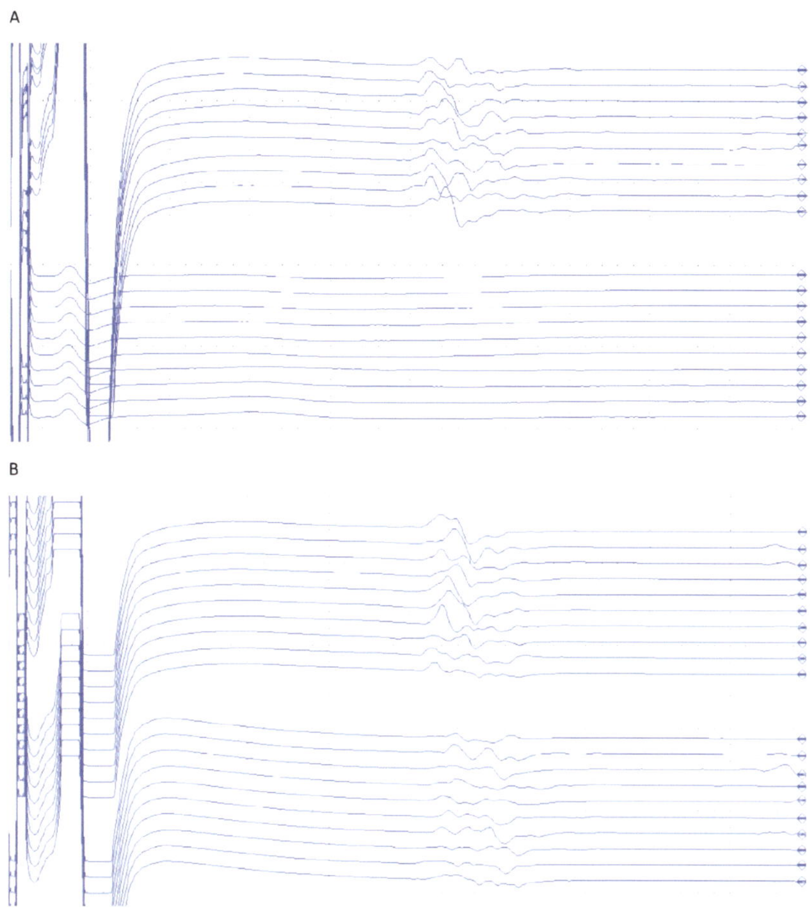

Figure 8.
F-responses of the tibial nerve in the abductor hallucis brevis muscle, latency 50 ms. stimulation at the ankle, upper 10 responses with surface electrodes. The lower 10 sweeps are recorded from the extensor digitorum brevis muscle: A) with a concentric needle electrode, no responses, and B) with surface electrodes, active electrode on the muscle and reference placed in the distal end of 5th metatarsal bone. Observe the "medium latency reflexes" of the extensor digitorum brevis muscle, which are volume conducted F-responses of the abductor hallucis brevis muscle. Calibration: 10 ms /div, 2 mV / div.

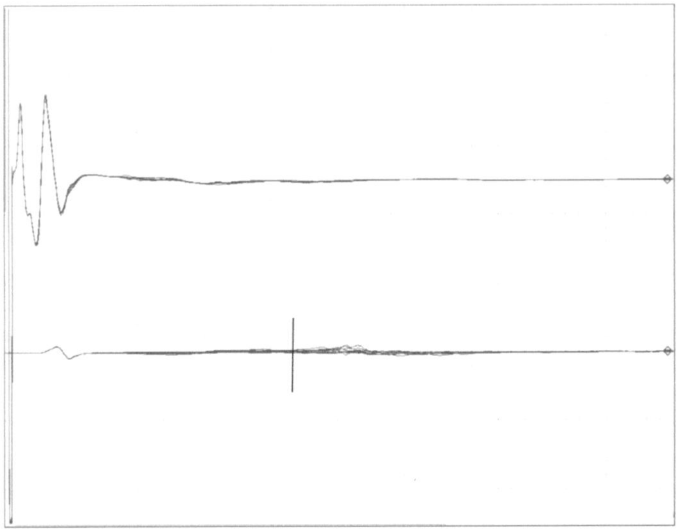

Figure 9.
*Stimulation of the common peroneal nerve at the fibular head. Uppermost: M-response of the anterior tibial muscle recorded with surface electrodes. Lowermost: Possible medium latency reflex responses of the abductor digiti minimi muscle, latency 85 ms (vertical line). But the volume conducted F-responses of the extensor digitorum brevis muscle might be another possibility (see **Figure 7B**). However, the F-response latency should be essentially shorter than this recorded response, with stimulation at the fibular head. Calibration 20 ms/div, 2 mV/div.*

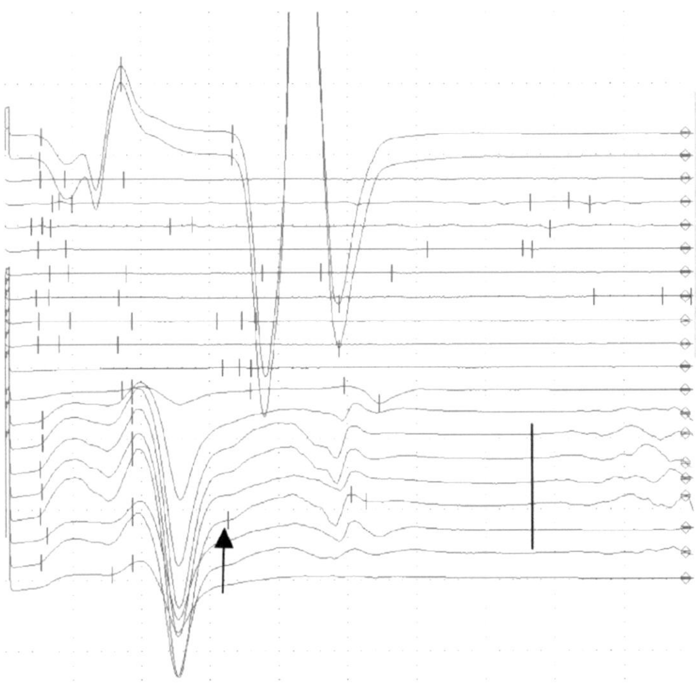

Figure 10.
H-reflex of the soleus muscle, latency 29.8 ms, stimulation of n. tibialis *at the popliteal space (uppermost sweeps). When the stimulation was changed to the common peroneal nerve (lowermost sweeps), "H-reflex" of the peroneus longus muscle was elicited (arrow). But we may also see a medium latency reflex response of about 64 ms (vertical line), which cannot be elicited from the peroneal muscles. This proves that a branch of the tibial nerve leading possibly to the lateral gastrocnemius muscle, with medium latency reflex response is also activated by the stimulation. Calibration 8 ms/div, 1 mV/div.*

DOI: http://dx.doi.org/10.5772/intechopen.95575

medium latency reflex response of the abductor digiti minimi muscle may really occur (**Figure 9**). The medium latency reflex response may reveal the false" H-reflex" of the peroneus longus muscle (**Figure 10**).

2. Conclusions

The H-reflex is a useful tool as a probe for clinical neurophysiologist, but the pathways and pitfalls should be considered [12]. The integrity of posterior roots of the S1 level is readily studied with the soleus H-reflex measurement, and we have used this method routinely for a long time. Unfortunately, there are difficulties with recording of H-reflex of the L5 level. Our trials with surface electrodes were often hampered by volume conducted reflex responses of the triceps surae muscle. With a simultaneous needle recording of the peroneus longus muscle we could prove that the H-reflex response of it is real, but it should be distinguished from F-responses. The medium latency reflex of the soleus muscle can be recorded with a powerful twitch contraction of the pretibial muscles elicited by supramaximal stimulation of the common peroneal nerve. We do not have any experience of its use in S1 root syndrome diagnostics. However, it is tempting to assume, that we may study the integrity of two completely different parallel proprioceptive reflex arches of the soleus muscle: the Ia-afferent – α efferent reflex arch (H-reflex), as well as the II-afferent – β efferent reflex arch (medium latency reflex), and compare the results of these measurements in investigation of the posterior root syndrome at S1 level. H-reflex of the peroneus longus muscle might be used in the diagnostics of posterior root lesion of the L5 level. We recommend the recording of this reflex with an EMG needle electrode. This recording can be performed accompanied with the needle EMG study for a possible axonal injury of the L5 motor nerve fibres of the peroneus longus muscle. L3 and L4 posterior roots might be investigated with the quadriceps femoris and adductor H-reflexes. Considering these many different methods and technical challenges related to them, a practitioner is advised to collect own normative data and always interpret the results rather conservatively in the context of full clinical picture.

Author details

Juhani Partanen[1]*, Urho Sompa[2] and Miguel Muñoz-Ruiz[2]

1 Department of Clinical Neurophysiology, University Central Hospital of Helsinki, Jorvi Hospital, Rörstrandinkatu 4 B 27, 00560, Helsinki, Finland

2 Department of Clinical Neurophysiology, Central Finland Health Care District, Keskussairaalantie 19, PL 123, 40620, Jyväskylä, Finland

*Address all correspondence to: junipartanen@gmail.com

References

[1] Partanen J. Different types of fibrillation potentials in human needle EMG. In: Turker H editor. Electrodiagnosis in new frontiers of clinical research. Rijeka, Croatia: Intech 2013, http://dx.doi.org/10.5772/55352

[2] Preston and Shapiro, editors. Electromyography and neuromuscular disorders: Clinical-electrophysiologic correlations. 2013. London: Elsevier.

[3] American Association of Neuromuscular & Electrodiagnostic Medicine. Don't perform dermatomal somatosensory evoked potentials (SEPs) for a pinched nerve in the neck or back, as they are an unproven diagnostic procedure. 2015, February 10.

[4] Burke D. Clinical uses of H reflexes of upper and lower limb muscles. Clinical Neurophysiology Practice. 2016;1:9-17.

[5] Partanen J. F-vaste ja H-refleksi. Chapter 5.1. In: Lang H, Partanen J, Häkkinen V, Larsen A. editors. Sähköiset hermomme. Turku: The Society of Clinical Neurophysiology in Finland 1992, p. 143-150.

[6] Ertekin C, Bademkiran F, Tataroglu C, Aydogdu I, Karapinars N. Adductor T and H reflexes in humans. Muscle & Nerve 2006;34:640-645.

[7] Burke D, Gandevia SC, McKeon B. The afferent volleys responsible for spinal proprioceptive reflexes in man. The Journal of Physiology 1983;339:535-552.

[8] Yusal H, Larsson L-E, Efendi H, Burke D, Ertekin C. Medium-latency reflex response of soleus elicited by peroneal nerve stimulation. Experimental Brain Research 2009;193:275-286.

[9] Wiesendanger M, Miles TS. Ascending pathway of low-threshold muscle afferents to the cerebral cortex and its possible role in motor control. Physiological Reviews 1982;62:1234-1269.

[10] Schieppati M, Nardone A, Siliotto R, Grasso M. Early and late stretch responses of human foot muscles induced by perturbation of stance. Experimental Brain Resarch 1995;105:411-422.

[11] Grey MJ, Ladouceur M, Andersen JB, Nielsen JB, Sinkjær T. Group II muscle afferents probably contribute to the medium latency soleus stretch reflex during walking in humans. The Journal of Physiology 2001;534.3:925-933.

[12] Knikou M. The H-reflex as a probe: Pathways and pitfalls. Journal of Neuroscience Methods 2008;17:1-12.

[13] Metso AJ, Palmu K, Partanen JV. Compound nerve conduction velocity - A reflection of proprioceptive afferents? Clinical Neurophysiology 2008;119:29-32.

[14] Nardone A, Schieppati M. Medium-latency response to muscle stretch in human lower limb: estimation of conduction velocity of group II fibres and central delay. Neuroscience Letters 1998;249:29-32.

[15] Kakuda N, Miwa T, Nagaoka M. Coupling between single muscle spindle afferents and EMG in human wrist extensor muscles: physiological evidence of skeletofusimotor (beta) innervation. Electroencephalography and Clinical Neurophysiology 1998;109:360-363.

Chapter 4

The Knee Proprioception as Patient-Dependent Outcome Measures within Surgical and Non-Surgical Interventions

Wangdo Kim

Abstract

Proprioception considered as the obtaining of information about one's own action does not necessarily depend on proprioceptors. At the knee joint, perceptual systems are active sets of organs designed to reach equilibrium through synergies. Many surgical procedures, such as ACL reconstruction in personalized medicine, are often based on native anatomy, which may not accurately reflect the proprioception between native musculoskeletal tissues and biomechanical artifacts. Taking an affordance-based approach to this type of "design" brings valuable new insights to bear in advancing the area of "evidence-based medicine (EBM)." EBM has become incorporated into many health care disciplines, including occupational therapy, physiotherapy, nursing, dentistry, and complementary medicine, among many others. The design process can be viewed in terms of action possibilities provided by the (biological) environment. In anterior crucial ligament (ACL) reconstruction, the design goal is to avoid ligament impingement while optimizing the placement of the tibial tunnel. Although in the current rationale for tibial tunnel placement, roof impingement is minimized to avoid a *negative* affordance, we show that tibial tunnel placement can rather aim to constrain the target bounds with respect to a *positive* affordance. We describe the steps for identifying the measurable invariants in the knee proprioception system and provide a mathematical framework for the outcome measure within the knee.

Keywords: knee proprioception, knee-tensegrity-structure (KTS), affordance-based-design, ACL impingement, knee synergy, entrainment, instantaneous knee screw (IKS)

1. Introduction

1.1 Anterior crucial ligament reconstruction and tibial tunnel placement

The anterior cruciate ligament (ACL) is a critical knee joint, bone-to-bone connected, stability ligament that is attached from an anterior location of the proximal tibia to a posterior location of the distal femur. The ACL is highly susceptible to failure during athletic activities and slip-fall events. The goal of ACL reconstruction surgery is to rebuild the ligament attachments as closely as possible to the native anatomy in order to restore pre-injury knee function and normal proprioception in the affected knee [1]. Personalized medicine in surgery allows the customization of insertion sites, graft size, tunnel placement, and graft tension for each individual patient [2]. A critical pre-operative decision concerns the placement of a tibial-femoral tunnel mimicking the native orientation of the ACL attachment [2]. Surgeons need to consider particular aspects of the local anatomy and, by extension, the biomechanical artifacts introduced during surgery. Here, we report an alternative approach based on the understanding of knee affordances to guide surgeons in the design of knee reconstruction strategies.

As aforementioned, an important predictor of clinical outcome during ACL reconstruction is tunnel placement [3, 4]. Roof impingement occurs when an ACL graft prematurely contacts the intercondylar roof before the knee reaches terminal extension. A tibial tunnel anterior to the tibial intersection of the intercondylar roof's slope allows the distal half of the roof to impinge on the anterior surface of the graft (*arrow I* in **Figure 1(a)**). Impingement syndrome occurs when the relationship between two articular components are incongruous, with resulting friction, inflammation, and degeneration [6]. Failure of grafts placed anteriorly is likely due to the impact of the bony roof on the graft's anterior surface during knee extension (**Figure 1(b)**) [5].

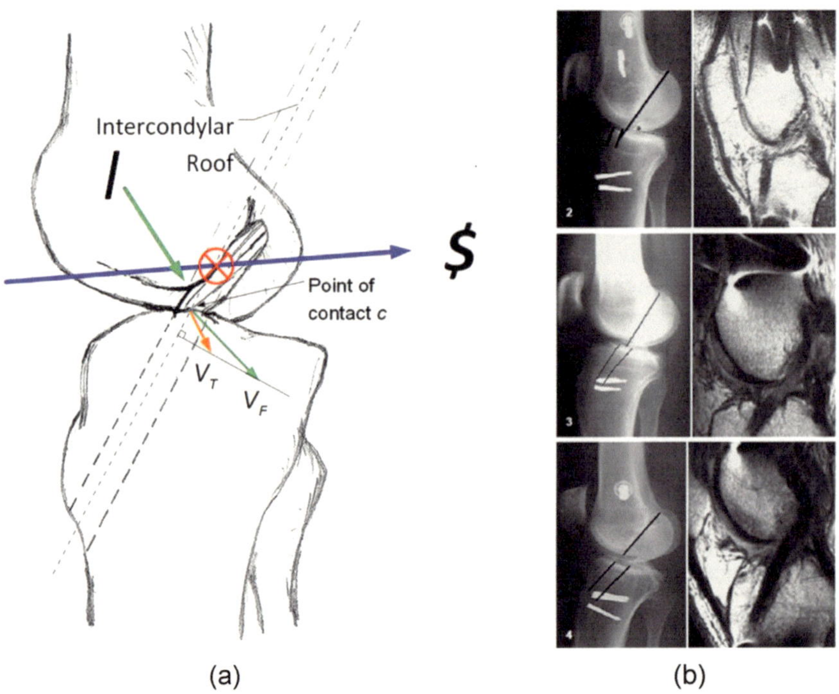

Figure 1.
(a) Schematic representation for the surgical placement of the tibial-femoral tunnel, presenting the possibility for uniform motion transmission within the knee. The (positive) affordance based *approach encourages surgical designers to customize the position of the tibial-femoral tunnel to intersect with the instantaneous knee screw (IKS or $). The point of contact (c) is determined from femoral and tibial velocity vectors during joint movement (V_F and V_B respectively). (b) Radiographic and magnetic resonance images (MRI) of an exemplar of tibia tunnel placement leading to roof impingement during ACL reconstruction.* Negative affordances *(severe and moderate impingement) inform about the tunnel locations to avoid, in order to prevent roof impingement of the ACL. Severe roof impingement occurs when the surgeon places the tibial tunnel in a local totally anterior to the slope of the intercondylar roof (top). Moderate roof impingement occurs when the surgeon places the tibial tunnel in a local partially anterior to the slope of the intercondylar roof (middle). A graft may also become impinged when the surgeon places the tibial tunnel in a local entirely posterior and parallel to the slope of the intercondylar roof (bottom). An impinged graft has a low, uniform signal intensity on the MRIs. The original schematic and images were published previously [5] and are used by the permission of Dr. Stephen Howell.*

To our knowledge, this is the first study to use psychological theory to address this surgical design concept [7]. Traditional rating systems to assess clinical outcome after joint arthroplasty are often based on the surgeon's objective ratings, such as range of motion and strength, or clinical ratings of function and pain. However, the patient's perceptions after arthroplasty may differ significantly from those of their clinician. Moreover, surgeons often underappreciate the needs and views of their patients [8]. There is, therefore, increasing awareness of the need to include patient-reported outcome (PRO) instruments in the evaluation of surgical procedures. Indeed, these patient-centered assessments of treatment outcomes are becoming today's standard [9]. Patient-reported outcome metrics (PROMs) can be simply described as a patient's health status self-report. A 'forgotten joint score', corresponding to when a patient forgets the artifact in their everyday life, was introduced in PROM as the ultimate goal in joint reconstruction [10]. 'Forgotten joint scores' are often observed in patients after surgery [11]. Nevertheless, these ratings do not replace the need to understand the general role of artifacts and affordances in reconstruction surgery. This study aims to identify measurable invariants using a (positive) affordance-based design strategy for structural tunnel placement during ACL reconstruction.

2. Affordance-based design

Current approaches in design science are characterized by a strong emphasis on methods as opposed to theory. Herbert Simon [12] was one of the early proponents studying design as a science. In the 1960s, Simon criticized the lack of a theoretical basis in design methods, describing such ad hoc methods as 'cook-book approaches.' Novel conceptual frameworks for design allow engineers to better describe and solve problems at the system level, such as those involving user interactions. We propose a conceptual approach for design based on affordances, a concept used in the study of perception in ecological psychology.

'Architecture and design do not have a satisfactory theoretical basis,' wrote psychologist James J. Gibson three decades ago. He also asked, 'can an ecological approach to the psychology of perception and behavior provide it?' [13]. Gibson's affordances theory describes how animals perceive their environment [14]. We applied Gibson's concept of affordance to the design of artifacts, in particular anatomic artifacts, which impacts on their biomechanics.

A decade after Gibson's seminal work, another psychologist, Donald A. Norman, use Gibson's theory of affordance to understand artifact design [15]. However, Norman's approach stopped short of incorporating the concept of affordance as fundamental to the design of any artifact [16]. When Norman revised the 1988 edition of his book in 2013, he rejected the ecological theory. He noted that the term affordance was often misused by psychologists, and as a result, he introduced the term 'signifier.' Signifiers make explicit that affordances are inputs used during cognitive deliberation for creating internal mental representations, which contradicts Gibson's claims that, if a designer successfully makes affordances possible, the artifact directly informs how it can be used—which is the hallmark of successful design. Intriguingly, in a recent study [17], Norman regrets that different psychology fields and design science have become separate silos unable to communicate with one another.

Ecological psychologist William Warren has applied the concept of affordances to the design of specific artifact-user relationships, such as the height of stairway steps [18]. His approach relied on the ratio of leg height to step height. Paola Cesari followed up on Warren's stair climbing studies by showing that older people perceive stairs differently than young people. However, the ratio between step height and the distance between the stepping foot and the top edge of the step was similar in both groups [19].

Since the concept of 'affordance' was introduced almost 40 years ago, it has been used in a variety of fields, including child psychology [20], the design of graphical user interfaces [21], mobile robots [22], control room interfaces [23], and more recently, in engineering design [24–26]. The impetus for any design project can be understood in terms of creating and changing affordances. The design process is the construction of an artifact that offers specific affordances, but not certain undesired affordances. An artifact with more positive affordances is considered better, while an artifact with more negative affordances is considered worse. However, this approach does not follow ecological psychology, but instead, it addresses the difficulty of identifying affordances with engineering [27].

Maier and Fadel coined the term artifact-artifact affordance (AAA) [24, 25]; however, AAA has not been properly incorporated within the larger theory of affordances. Although AAA was developed as a new concept, the idea that inanimate objects offer action possibilities in an organism is a foundational concept known since Gibson's work in ecological psychology. The ecological approach demonstrates how animal (including human) perception and action is continuous with interactions between inanimate physical systems, or the world in general. The entrainment of separate limbs during biological coordination, for example, follows the same physical laws as entrainment between two pendulum clocks or other purely mechanical (inanimate) systems [28]. The fact that interactions between inanimate and animate systems are continuous precludes the need to identify AAA as a distinct category.

In short, these concepts should be used with great care if knowledge is to be gathered. In the present study, we used a surgical technique as an example of how the theory of affordances may be utilized for affordance-based design.

3. Artifact-user affordances versus artifact-artifact affordances

Gibson demonstrated how animal perception and action is continuous, with interactions with inanimate objects or surfaces [14]. The affordances of a product are what it provides, offers, or furnishes to a user. Gibson's 'system theory' of perception corresponds to an open system, which is rather different from the view of isolated artifacts [29]. For engineering design, an affordance can be defined as the relationship between person and artifact from which the behavior emerges. These affordances between artifacts and the people that use them are called artifact-user affordances (AUA).

For example, the gear pair (**Figure 2(a)**) is referred to as an artifact-artifact affordance (AAA) for uniform motion transmission between two parallel axes, and it is possible only if the line of action passes through a fixed point, known as the pitch point. Moreover, assuming that gear 1 rotates with constant angular velocity ω_1, the motion is transferred by direct contact at points K1 and K2. The objective is to determine whether or not the angular velocity ω_2 will remain constant or present uniform motion transmission. Kennedy's theorem identifies the fundamental property of two interacting rigid bodies in motion [30], such that three instantaneous centers shared by three rigid bodies in relative motion to one another, all lie on the same straight line. Uniform motion transmission between two parallel axes is possible only if the line of action passes through an invariant point, known as the pitch point. The pitch point is the instantaneous center of velocity for the two gears. For the gear teeth to remain in contact, the two-component velocities along the common normal must be equal. The absolute velocities along the line of action must

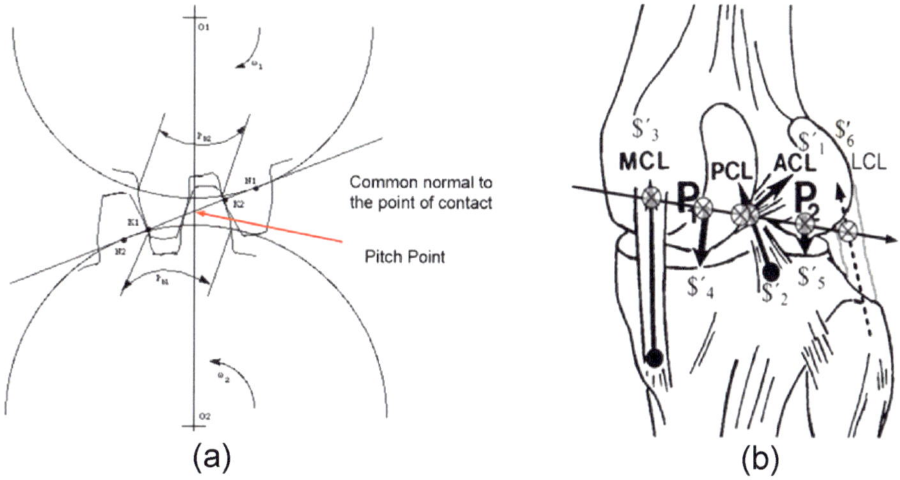

Figure 2.
(a) The artifact-artifact conjugate action, in the form of two interacting gears, demonstrating the uniform motion transmission between two parallel axes as can be found in a knee joint. (b) the knee joint synergy as represented by six constraints $\left(\$'_i, i = 1,..,6\right)$*, which are conjointly reciprocal to the instantaneous knee screw ($) as indicated by their intersections (at the ⊗ 's). A balance of forces happens when the virtual coefficient vanishes, being it the necessary and sufficient condition for knee equilibrium.*

be identical; otherwise, bodies 1 and 2 become separated. It was shown that if the involute profile describes *the gear profile*, the common normal does not change its direction because it is an invariant of the structure. However, this may not be the case if we consider it in terms of affordances.

Gibson claims that some affordances are beneficial while others are injurious, such as maintaining a line of action versus veering off course [14]. These benefits and disadvantages, safeties and dangers, positive and negative affordances are properties of events taken with reference to an observer, and not properties of the observer's experiences; they are not personal values or feelings of pleasure or pain added to neutral perceptions [14]. For example, physical properties of the tibial tunnel and the intercondylar notch roof are not affordances in and of themselves, but they do determine what affordances are offered to a surgeon depending on a patient's anatomic features. Thus, the characteristics that affect positive AUA are the same as those affecting negative AUAs. The artifact only has one set of characteristics, which is a customization of the tunnel placement, and this is all that the designers or surgeons can act upon. As a consequence of such mutuality, affordances do not exist in the patient's tunnel or intercondylar roof, but in what they offer to the surgeon. Importantly, AUAs may conflict with one another when the graft becomes slack or loose (i.e., loss of extension in the graft at full extension, or the graft being trapped in the notch), indicating a negative affordance or an increase in the potential for injury. Thus, when a surgical designer identifies a functional range in which a joint is not allowed to fail, they need to constrain the target bounds for that same joint to enhance positive affordances and avoid negative affordances. This approach is addressed in the section below.

4. The affordance-based design applied to reconstructed knee joint function

A joint 'gear' cannot perceive itself or its joint gear since gears are inanimate. Gears simply conjugate uniform motion transmission by virtue of their tensegrity structure, manifesting that structure influences behavior [31]. However, the knee is an active set of bone structures that come to equilibrium via a joint function. The function of a joint is not only to permit mobility of the articulated bones but also to maintain a stable bone position and movement. Knee structures include muscles/tendons, anterior cruciate ligaments (ACL), posterior cruciate ligament (PCL), medial collateral ligament (MCL), lateral collateral ligament (LCL), and articular cartilage contact in the medial (P_1) and lateral (P_2) compartments (**Figure 2(b)**). From a biomechanical point of view, several studies have shown that these intra-articular knee structures work in synergy with the ACL [32–34]. The knee synergy engages afferent/efferent motor control loops that establish functional equilibrium gait patterns [35].

Neurophysiologist Nikolai Bernstein defined coordination as mastering the many degrees of freedom (DOF) of a particular movement by reducing the number of variables to be controlled [36]. Recently, a contemporary perspective on Bernstein's concept of synergies has been proposed [37]. The muscle synergy is equivalent to the complexity of lines, a manifold approximated by individual fibers (**Figure 3(a)**). Muscles are not functional units, even though this is a common misconception. Instead, most muscular movements are generated by many individual motor units distributed over some portions of one muscle, plus portions of other muscles. The tensional forces of these motor units are then transmitted to a complex network of fascia sheets, bags, and strings, which convert them into the final joint/body movement [38].

Line manifold contraction is a linear line complex [39] defined by screws ($IS(\mathbf{p})$) (**Figure 3(b)**). Bodies twist around a screw, called Instantaneous Screw [40]. In any screw motion along a line axis forming a linear complex, the lines remain within the complex. Additional cognitive processes or internal representations are not needed to explain these phenomena, as perception and action are coupled. Perceptual systems are active sets of organs designed to reach equilibrium through synergies [41]. A body

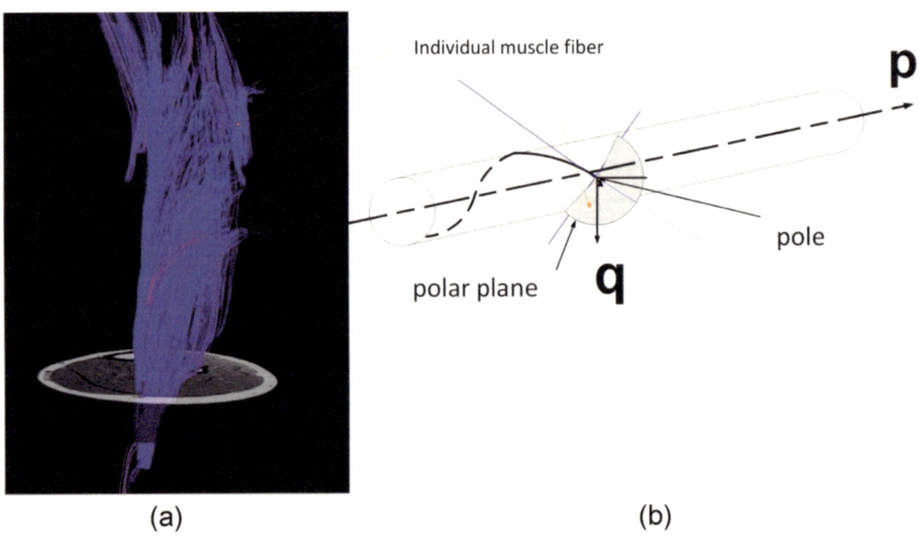

Figure 3.
(a) Fiber tractography image of a portion of the lateral gastrocnemius muscle as demonstrated in an exemplar healthy subject. The overlying images were generated at one region of interest, corresponding to the muscle boundaries where the anatomical cross-section area was maximal (whole-body MRI scanner, Signa HDxT 1.5 T, GE Healthcare, USA). The patients were placed in the supine position, feet first, and the position of each participant was considered in relation to the long axis of the leg, which was placed in parallel to the magnetic field. (b) a manifold of muscle fibers in tension forms to the linear complex identified as an instantaneous screw $IS(\mathbf{p})$ *and its perpendicular pole* ($\mathbf{q}$) *within the synergy of gait.*

cannot remain in equilibrium if the fiber forces that act upon the body have a non-balance resultant force. The velocity vector of every point in the fiber segment is tangential to the helix passing through it. The pattern of this velocity vector is a helicoidal velocity field. Each point that does not coincide with the screw's twist $IS(\mathbf{p})$ is referred to as a pole ($\mathbf{q}$). Associated with each pole is its corresponding polar plane. A polar plane and its corresponding pole, as defined by the instantaneous screw, have been illustrated here (**Figure 3(b)**).

In our previous research [42, 43], we introduced the concept of measurable invariant of the knee perceptual organ. In such invariant, six constraints ($) are collectively reciprocal to the instantaneous knee screw (IKS or $) indicated by ⊗ (**Figure 2(b)**). These metrics predicted the knee synergy model based on synergies [44]. Moreover, this perspective defines torque-free pure forces based on the tensegrity structure [45–48]. It is important to note that this configuration is a tensegrity configuration, as the system is pre-stressable in the absence of external forces, such as ground reaction forces during actual locomotion [49]. It was shown the knee tensegrity structure (KTS) has six constraints, and that it can balance the forces between tension and compression in the joint such that no work results

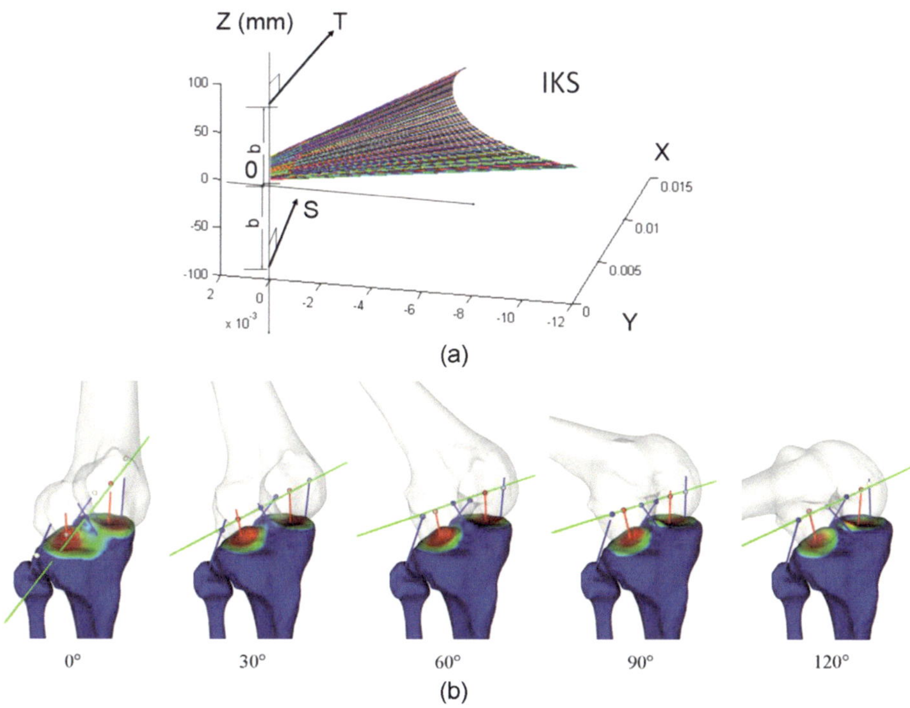

Figure 4.
(a) An exemplar Ball-Disteli diagram [52] with two generally disposed screws (T and S or $\mathbf{p}_1$ and $\mathbf{p}_2$), conveniently placed on the z-axis along their common perpendicular (with b). The origin of coordinate O is halfway between the screws, and the x-axis is inclined by half the included angle σ between the two screws (S, T). An instantaneous knee screw (IKS) after normalization is linearly dependent on the screws during any point of knee movement. The Ball-Disteli diagram aligns itself using the principle of three axes. (b) Representation of the IKS (green line); see Online Supplementary Video 1 (video is available via the following link: https://drive.google.com/file/d/18_YtszzT3_IvNIken5uxObj4jmSd0Zs_/view?usp=sharing). The lines of action of the ligaments (blue lines) and cartilage contact (red areas and red lines) for wrenches identified every 30° within the range of knee motion for a given patient. The white or colored dots represent the closest point to the IKS for each wrench. The colors range from white (di > 5 mm) to the color of the corresponding intersecting line (di = 0 mm). Cartilage contact colors on the tibia are proportional to the tibial-femoral relative separation (red: Distance ≤0 mm; blue: Distance >7 mm). The original anatomic schematics and lines of action were published previously [51] and are used by permission of professor Michele Conconi.

[50]. The KTS can be pre-stressed to obtain the same configuration as if external loads were applied. The selected pre-stress may yield the same configuration in the swing phase (external forces are absent) as in the stance phase (external forces are present) [49]. Notably, preparedness is not only a reactive aspect of the movement apparatus, but it also relates to anticipatory adjustments that predispose a system to behave in a particular way [37].

If a knee joint is only free to twist about a screw IKS while in equilibrium, despite being acted upon by the fiber reaction, the mechanical work during a small displacement against the reaction forces $\$'$ in the KTS must be zero, according to the following relationship [40],

$$\$^{T} \cdot KTS = 0 \tag{1}$$

Uniform motion transmission between two axes (defining the thigh and shank, respectively) is affordable only if their lines of action pass through the IKS, as expressed in the Eq. (1). Thus, the affordances of the knee synergy must be positive, and the joint ligaments should remain in an isometric/isokinetic condition or continuous length/tension. If not, the ligaments become slack or loose, resulting in roof impingement, post-reconstruction [5]. Moreover, Eq. (1) also implies that the moving self ($) and the invariant structure of the KTS reaction are reciprocal aspects of the same perception. Gibson called this information gathering approach propriospecific, as opposed to exterospecific, to specify the observer (here the self) as distinguished from the environment.

The knee synergy approach proposed herein was recently validated experimentally [51]. The authors calculated if all the lines of action intersect at the IKS ($) following natural knee motion to describe the knee surgery invariant. The results show the mean distances between each constraint line of action, and the IKS stayed below 3.4 mm and 4.5 mm for *ex vivo* and *in vivo* assessments, respectively (**Figure 4(b)**).

5. The affordance-based design applied to graft placements during reconstruction

It has been hypothesized that a tensegrity system serves the medium of haptic perception, from the individual cells to the whole body, maintaining continuous tension and discontinuous compression [53], which clearly exhibit the determinate character of the entire body system perception [38, 40]. In this study, we present the positive affordance-based design on graft placement while in continuous tension, rather than designing against the negative affordance by preventing impingement. As described, we use the invariant structure of the KTS [54] as an appropriate ecological frame of reference to locate the tibial tunnel placement. For the ACL-patient to engage the IKS directly, clinicians have to measure the tunnel placement relative to the posture and behavior of the person being considered, making continuous graft tension possible. First, an invariant should not be applied to the patient directly, for it is not a stimulus. Second, invariants can be considered qualitative rather than quantitative so that other clinical assessments can make it available to their surgeons/observers in an exact mathematical description [14].

The IKS is defined in terms of the second-order invariant by a linear combination of the two screws of the first-order invariant, S and T, instantaneous screw axes of the shank and thigh (**Figure 4(a)**) [55]. Then the IKS must be a screw that has been picked up from the many candidate screws on the cylindroids [40], which is reciprocal to KTS (via Eq. 1). Hence, the ratio of the amplitudes about S and T

may be determined (**Figure 4(a)**), which manifests the fact that the sensitivity of the knee joint to its disposition is of crucial importance in picking up information.

Two lines were projected respectively to the sagittal plane so that the path of the graft could be aligned to any transversal axis intersecting the IKS ($), the central line of KTS that is the second-order invariant line, also called the IKS (**Figure 1(a)**). The lines were generated at full knee joint extension. Notice that if the graft line is not precisely aligned with the member line within the KTS, due to position errors, for example, the velocity difference on the graft line would not be zero, but would still be small. If the path of an ACL graft is so selected that it cuts the IKS of the KTS, then the line becomes a member of the KTS, which ensures the isokinetic graft placement related to trans-tibial-femoral tunneling. Consider now the necessary kinematic relations in that contact point c as the common point belonging to both the tibial and femoral tunnels (**Figure 1(a)**).

The velocity of the point c residing on the femoral tunnel (V_F) can be resolved into two components: one component is perpendicular to the graft line and the other element parallel to it. Similarly, the velocity of the point on the tibial tunnel coincident with point c (V_T) can also be resolved into two components. For the two bony bodies (femur and tibia) to remain through one continuous body, the parallel component of the graft line for velocity must be equal, by projecting V_F and V_T onto the graft line (**Figure 1(a)**). The graft without that qualification would experience impingements. The difference in the perpendicular component represents the relative transverse velocities between the articulating tunnels and is closely related to an essential factor in choosing the proper tunnel width. Widening of the tunnel diameter might be performed, allowing more tolerance for this transverse velocity relationship, taking into account the width of the graft and the existing diameter of the notch.

As described, we identified the measurable second-order invariant of knee synergy and proposed it as a new view of the basis of tunnel placement by using Eq. (1). The knee synergy approach identifies the information as a means to perceive the affordance of uniform motion transmission. To apply the described approach and identify the invariant, we characterized the shank to the thigh (the tibia to the femur) relative motion, i.e., the second-order invariance of the knee synergy. These results were then compared with experimental data for validation as provided by the "Grand Challenge Competition to Predict *In Vivo* Knee Loads" as part of the Symbiosis project funded by the National Institutes of Health [56].

6. Entrainment of touch and posture

Contrasting the established idea of senses, Gibson considered separate anatomical units as perceptual systems [29]. In the present case, a joint yields spatial information, skin-nerve conveys contact information, and in certain dynamic combinations, joint and skin-nerve yield synchronization, or entrainment specifying information about the layout of external surfaces during locomotion.

Behavioral dynamics in a consistent approach has proposed to account for the dynamics of perception and action [57]. This approach followed Gibson's idea that rather than being localized in an internal (or external) structure, control is distributed over the agent-environment system, in the present case, the user-artifact-surface system. Therefore, Warren's behavioral dynamics argues for a one-to-one correspondence between the internal structure IKS, constituted by the internal forces formed by the distal end of the femur and the proximal end of the tibia, and the external structure, represented by the ground reaction forces (GRFs) on foot [58].

Behavioral dynamics control laws indicate that the entrainment or coordination of shank and thigh (S, T) follows the same physical laws as the entrainment between the knee and ground (IKS, GRF). Therefore, the cross-ratio [59] of the ordered pair (IKS, GRF) with respect to the ordered pair (S, T) is

$$\{(IKS,GRF);(S,T)\} = -1. \tag{2}$$

For a given IKS (when an observer perceives the affordance of the surface) and the location of the center of pressure (COP) on the axis of the GRF is known, then the GRF vector is limited to a plane in the screw system of the first order [47, 48] (**Figure 5(a)**). The muscle synergy η and GRF φ are then compounded into an invariant, limited to the plane of the COP in reciprocity with the IKS. This theorem was originally proposed by Möbius, who showed that forces from six lines could be equilibrated, and also, if five of the lines are given along with a point on the sixth line, then the sixth line is limited to a polar plane [40].

To test such ecological approach to perception and action during the stance phase of a gait, we compared previously published experimental data sets [56] with our predicted datasets [47, 48] in terms of medial and lateral contact forces. Available data included limb motion capture, fluoroscopy images, GRFs, electromyographical readings determining muscle forces, as well as medial and lateral knee contact forces derived from GRFs. Data were collected from an adult male with a right knee reconstruction (65 kg mass and 1.7 m height). When the variations in the ground contact (magnitudes and direction) were shown along with the variations of knee movement in terms of IKS, an invariant was determined uniquely by the two corresponding pairs, see Eq. (2) (**Figure 5(b)**).

In this study, the IKS was determined by a linear combination of two instantaneous screw axes of the shank and thigh (**Figure 4(a)**). The IKS nearly coincides with a reciprocal screw of the GRF, as indicated in a magnified inset image in **Figure 5(b)**. A perceptual system of the knee can come to equilibrium since twists of amplitudes S and T neutralize. We thus see that the evanescence of one

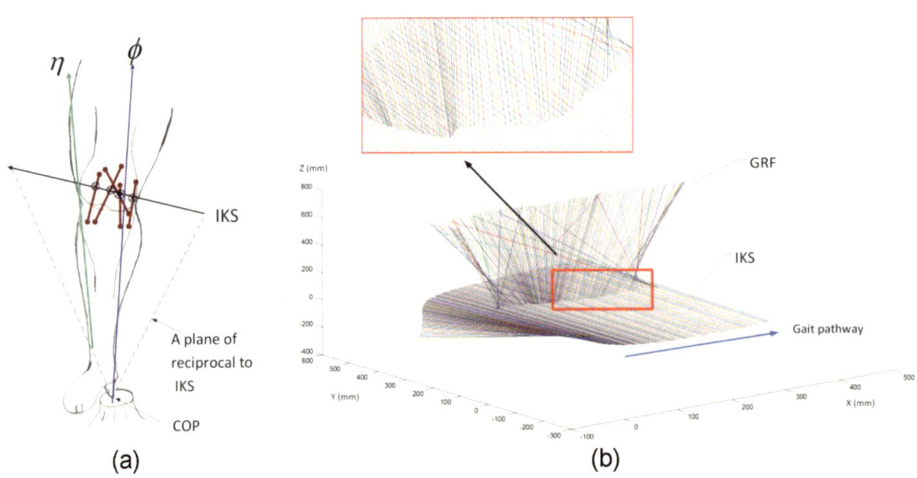

Figure 5.
(a) The framework for estimating responses to constraints on the knee joint (ligament forces and contact forces) is influenced by the inclusion of muscle synergy (η) and GRF (φ) relative to the center of pressure (COP). The judicious generation of the IKS for the one DOF in knee equilibrium simplifies the estimation. This figure was adapted from the original figure published previously [47, 48]. (b) Perception and action during the stance phase of gait entrain the knee joint rotation with the touch pattern (GRF) of the foot. The invariant knee-manifolds demonstrates that an affordance for postural stability is measured relative to the posture of the patient, as represented by the entrainment of the GRF with the IKS at any point in the gait pathway.

function must afford all that is necessary for subordinate organs (S, T) belong to an IKS of the superordinate organ for information pickup over paths of locomotion. This reciprocity is captured by the concept of a mutual relationship between the constraints and the DOF [60, 61]. Information about the person accompanies information about the environment. Here it is shown that proprioception accompanies exteroception; information is available to specify both poles [14].

A lateral radiograph of the knee in extension was the traditional approach to diagnose any roof impingement, and a portion of the tibial tunnel was traditionally placed anterior to the intercondylar roof [5] (**Figure 1(b)**). However, the available information on the experimental images can not be applied to another patient because they do not provide environmental information. Thus, AUA has the potential to diagnose pathologies. The last decade has seen a paradigm shift in the measurement of clinical outcomes, with an increasing focus on the user's perspective, PROMs. Many clinicians, though, are less confident in self-reported PROMs, than in 'objective measurements' [11]. Recent studies identified several sensations, activities, and psychological factors such as feelings of instability and knee-related fears that make the patients aware of their artificial knee joint [62]. They concluded that joint awareness might work as an overarching parameter. This is aligned with Gibson's statement that an affordance cuts across the dichotomy of subjective-objective and helps us to understand its inadequacy [14]. Affordances have to be designed in relation to the uniqueness of each patient, and thus posture and movement need to be measured in terms of a specific patient-environment system, not in patient-centered terms.

7. Conclusion

This study presented an affordance based design supporting knee reconstruction surgery, with applications to the user/surgeon/therapist. It brings ecological theory to robustly explain knee biomechanics and clarifying the general role of physical artifacts and affordances in surgery. The mutuality of user and artifact that we defended here is not traditionally guiding individualized ACL reconstruction. Instead, the anatomic ACL reconstruction seems to lead to the idea that a deficient ACL is not understandable within knee joint biomechanics [32]. As we argued, the ACL is a highly organized synergy with intra- and extra-articular components [34] and yet still an identifiable system within the anatomic environment. The knee complexes in Eq. (2) reinforce how perception and action are coupled. A unique combination of invariants, a compound invariant, is just another invariant [14]. In particular, this study identified the knee complexes as the measurable invariable structures that specify the persisting placement of the tunnel during ACL reconstruction.

Acknowledgements

Author WK extends thanks to Ms. Flávia Yázigi for her hard work with the radiography and a long recruitment process. WK also thanks to his mother-in-law, Ms. Sun Lee, for her continuous encouragement for this research. The experimental data used for validation were provided by the "Grand Challenge Competition to Predict *In Vivo* Knee Loads" as part of the Symbiosis project funded by the US National Institutes of Health via the NIH Roadmap for Medical Research (Grant # U54 GM072970). WK also thanks Dr. Michele Conconi of University of Bologna, for making the videos "The Geometrical Arrangement of Knee Constraints That Makes Natural Motion Possible" available to us.

Author details

Wangdo Kim
Ingeniería Mecánica, Universidad de Ingenieria y Tecnologia – UTEC, Lima, Perú

*Address all correspondence to: mwdkim@utec.edu.pe

References

[1] Behrend, H., K. Giesinger, V. Zdravkovic and J. M. Giesinger (2017). "Validating the forgotten joint score-12 in patients after ACL reconstruction." The Knee **24**(4): 768-774.

[2] Karlsson, J., M. T. Hirschmann, R. Becker and V. Musahl (2015). "Individualized ACL surgery." Knee Surgery, Sports Traumatology, Arthroscopy **23**(8): 2143-2144.

[3] Howell, S. M. and M. L. Hull (2009). "Checkpoints for judging tunnel and anterior cruciate ligament graft placement." J Knee Surg **22**(2): 161-170.

[4] Scheffel, P. T., H. B. Henninger and R. T. Burks (2013). "Relationship of the intercondylar roof and the tibial footprint of the ACL: implications for ACL reconstruction." Am J Sports Med **41**(2): 396-401.

[5] Howell, S. M. (1998). "Principles for placing the tibial tunnel and avoiding roof impingement during reconstruction of a torn anterior cruciate ligament." Knee Surgery, Sports Traumatology, Arthroscopy **6**(1): S49-S55.

[6] Faletti, C., N. De Stefano, G. Giudice and M. Larciprete (1998). "Knee impingement syndromes." Eur J Radiol **27 Suppl 1**: S60–S69.

[7] Niama Natta, D. D., E. Thienpont, A. Bredin, G. Salaun and C. Detrembleur (2019). "Rasch analysis of the Forgotten Joint Score in patients undergoing knee arthroplasty." Knee Surgery, Sports Traumatology, Arthroscopy **27**(6): 1984-1991.

[8] Kinnaman, J. E. S., A. D. Farrell and S. W. Bisconer (2006). "Evaluation of the Computerized Assessment System for Psychotherapy Evaluation and Research (CASPER) as a Measure of Treatment Effectiveness With Psychiatric Inpatients." Assessment **13**(2): 154-167.

[9] Rolfson, O., K. Eresian Chenok, E. Bohm, A. Lübbeke, G. Denissen, J. Dunn, S. Lyman, P. Franklin, M. Dunbar, S. Overgaard, G. Garellick, J. Dawson and R. Patient-Reported Outcome Measures Working Group of the International Society of Arthroplasty (2016). "Patient-reported outcome measures in arthroplasty registries." Acta orthopaedica **87 Suppl 1**: 3-8.

[10] Behrend, H., K. Giesinger, J. M. Giesinger and M. S. Kuster (2012). "The "Forgotten Joint" as the Ultimate Goal in Joint Arthroplasty." The Journal of Arthroplasty $V 27(3): 430-436.e431.

[11] Hamilton, D. F., J. M. Giesinger and K. Giesinger (2017). "It is merely subjective opinion that patient-reported outcome measures are not objective tools." Bone & joint research **6**(12): 665-666.

[12] Simon, H. A. (2008). The sciences of the artificial.

[13] Gibson, J. J. (1976). "The Myth of Passive Perception: A Reply to Richards." Philosophy and Phenomenological Research **37**(2): 234-238.

[14] Gibson, J. J. (1979). The ecological approach to visual perception, Houghton Mifflin.

[15] Norman, D. A. (1988). The Psychology of Everyday Things, Basic Books.

[16] Norman, D. A. (2013). The Design of Everyday Things, Doubleday.

[17] Norman, D. A. (2015). "Affordances: Commentary on the Special Issue of AI EDAM." Artificial Intelligence for

Engineering Design, Analysis and Manufacturing **29**(3): 235-238.

[18] Warren, W. H., Jr. (1984). "Perceiving affordances: visual guidance of stair climbing." J Exp Psychol Hum Percept Perform **10**(5): 683-703.

[19] Cesari, P. (2005). "An invariant guiding stair descent by young and old adults." Exp Aging Res **31**(4): 441-455.

[20] Gibson, E. J., E. J. G. A. D. Pick, A. D. Pick, c. 020321/l, P. I. C. D. A. D. Pick and S. L. S. P. P. E. J. Gibson (2000). An Ecological Approach to Perceptual Learning and Development, Oxford University Press.

[21] Cairns, P. and H. Thimbleby (2008). "Affordance and Symmetry in User Interfaces." The Computer Journal $V 51(6): 650-661.

[22] Iagnemma, K. and J. Overholt (2015). "An Architecture for Online Affordance-based Perception and Whole-body Planning." Journal of Field Robotics $V 32(2): 229-254.

[23] Vicente, K. J. and J. Rasmussen (1990). "The Ecology of Human-Machine Systems II: Mediating 'Direct Perception' in Complex Work Domains." Ecological Psychology **2**(3): 207-249.

[24] Maier, J. R. A. and G. M. Fadel (2009a). "Affordance-based design methods for innovative design, redesign and reverse engineering." Research in Engineering Design **20**(4): 225.

[25] Maier, J. R. A. and G. M. Fadel (2009b). "Affordance based design: a relational theory for design." Research in Engineering Design **20**(1): 13-27.

[26] Maier, J. R. A., G. M. Fadel and D. G. Battisto (2009). "An affordance-based approach to architectural theory, design, and practice." Design Studies **30**(4): 393-414.

[27] Maier, J. R. A. and G. M. Fadel (2007). Identifying affordances. 14th Int. Conf. Engineering Design (ICED07), Paris.

[28] Kelso, J. A. S. (1995). Dynamic patterns : The self-organization of brain and behavior. Cambridge, Mass., MIT Press.

[29] Gibson, J. J. (1966). The senses considered as perceptual systems. Boston, Houghton.

[30] Dooner, D. and A. Seireg (1995). The kinematic geometry of gearing: a concurrent engineering approach, Wiley-Interscience Publication.

[31] Forrester, J. W. (1990). Principles of systems, Productivity Press.

[32] Wroble, R. R., E. S. Grood, J. S. Cummings, J. M. Henderson and F. R. Noyes (1993). "The role of the lateral extraarticular restraints in the anterior cruciate ligament-deficient knee." The American Journal of Sports Medicine **21**(2): 257-263.

[33] Lane, J. G., S. E. Irby, K. Kaufman, C. Rangger and D. M. Daniel (1994). "The Anterior Cruciate Ligament in Controlling Axial Rotation:An Evaluation of Its Effect." The American Journal of Sports Medicine **22**(2): 289-293.

[34] Sonnery-Cottet, B., M. Thaunat, B. Freychet, B. H. B. Pupim, C. G. Murphy and S. Claes (2015). "Outcome of a Combined Anterior Cruciate Ligament and Anterolateral Ligament Reconstruction Technique With a Minimum 2-Year Follow-up." The American Journal of Sports Medicine **43**(7): 1598-1605.

[35] Blake, R. (1994). "Gibson's inspired but latent prelude to visual motion perception." Psychol Rev **101**(2): 324-328.

[36] Turvey, M. T. (1990). "Coordination." American Psychologist **45**(8): 938-953.

[37] Profeta, V. L. S. and M. T. Turvey (2018). "Bernstein's levels of movement construction: A contemporary perspective." Human Movement Science **57**: 111-133.

[38] Myers, T. W. (2001). Anatomy trains: Myofascial meridians for manual and movement therapists. Edinburgh ; New York, Churchill Livingstone.

[39] Jessop, C. M. (1903). Treatise on the Line Complex, American Mathematical Society.

[40] Ball, R. (1900). A treatise on the theory of screws, Cambridge University Press.

[41] Smart, B. M. (1988). Perception Without Processing [microform] : J.J. Gibson's Ecological Approach, Thesis (M.A.)--University of British Columbia.

[42] Kim, W., M. Espanha, A. Veloso, D. Araújo and F. João (2013a). "An Informational Algorithm as the Basis for Perception-Action Control of the Instantaneous Axes of the Knee." J Nov Physiother **3**(127): 2.

[43] Kim, W., Y.-H. Kim, A. P. Veloso and S. S. Kohles (2013b). "Tracking knee joint functional axes through Tikhonov filtering and Plűcker coordinates." Journal of novel physiotherapies(1).

[44] Turvey, M. T., H. L. Fitch and B. Tuller (2014). The Bernstein Perspective: I. The Problems of Degrees of Freedom and Context-Conditioned Variability. Human motor behavior: An introduction. J. S. Kelso, Psychology Press.

[45] Kim, W. and S. S. Kohles (2012). "A reciprocal connection factor for assessing knee-joint function." Computer Methods in Biomechanics and Biomedical Engineering **15**(9): 911-917.

[46] Kim, W., A. Veloso, J. Tan and C. Andrade (2010). A Reciprocal Connection at Knee Joint. ASME 2010 Summer Bioengineering Conference, Naples, FL.

[47] Kim, W., A. P. Veloso, D. Araújo, V. Vleck and F. João (2013c). "An informational framework to predict reaction of constraints using a reciprocally connected knee model." Computer Methods in Biomechanics and Biomedical Engineering: 1-12.

[48] Kim, W., A. P. Veloso, V. E. Vleck, C. Andrade and S. S. Kohles (2013d). "The stationary configuration of the knee." J Am Podiatr Med Assoc **103**(2): 126-135.

[49] Skelton, R. E. and M. C. d. Oliveira (2009). Tensegrity systems. Dordrecht ; New York, Springer.

[50] Huang, C., W. Kuo and B. Ravani (2008). On the Linear Line Complex and Helicoidal Vector Field Associated with Homologous Lines of a Finite Displacement.

[51] Conconi, M., N. Sancisi and V. Parenti-Castelli (2019). "The Geometrical Arrangement of Knee Constraints That Makes Natural Motion Possible: Theoretical and Experimental Analysis." Journal of Biomechanical Engineering **141**(5): 051001-051001-051006.

[52] Figliolini, G., H. Stachel and J. Angeles (2007). "A new look at the Ball–Disteli diagram and its relevance to spatial gearing." Mechanism and Machine Theory **42**(10): 1362-1375.

[53] Turvey, M. T. and S. T. Fonseca (2014). "The medium of haptic perception: a tensegrity hypothesis." J Mot Behav **46**(3): 143-187.

[54] Kelso, J. A. S. and G. Schöner (1988). "Self-organization of coordinative movement patterns." Human Movement Science 7(1): 27-46.

[55] Dooner, D. B. (2002). "On the Three Laws of Gearing." Journal of Mechanical Design **124**(4): 733-744.

[56] Fregly, B. J., T. F. Besier, D. G. Lloyd, S. L. Delp, S. A. Banks, M. G. Pandy and D. D. D'Lima (2012). "Grand challenge competition to predict in vivo knee loads." Journal of Orthopaedic Research **30**(4): 503-513.

[57] Warren, W. H. (2006). "The dynamics of perception and action." Psychological Review **113**(2): 358-389.

[58] Beer, F. P. (2010). Vector Mechanics for Engineers: Statics and dynamics, McGraw-Hill Companies.

[59] Semple, J. G. and G. T. Kneebone (1960). Algebraic Projective Geometry, Clarendon Press.

[60] Riley, M. A. and M.-V. Santana (2000). "Mutuality Relations, Observation, and Intentional Constraints." Ecological Psychology **12**(1): 79-85.

[61] Wagman, J. B. and C. Carello (2001). "Affordances and Inertial Constraints on Tool Use." Ecological Psychology **13**(3): 173-195.

[62] Loth, F. L., M. C. Liebensteiner, J. M. Giesinger, K. Giesinger, H. R. Bliem and B. Holzner (2018). "What makes patients aware of their artificial knee joint?" BMC Musculoskeletal Disorders **19**(1): 5.

Chapter 5

Proprioceptors in Cephalic Muscles

Juan L. Cobo, Sonsoles Junquera, José Martín-Cruces, Antonio Solé-Magdalena, Olivia García-Suárez and Teresa Cobo

Abstract

The proprioception from the head is mainly mediated via the trigeminal nerve and originates from special sensitive receptors located within muscles called proprioceptors. Only muscles innervated by the trigeminal nerve, and rarely some muscles supplied by the facial nerve, contain typical proprioceptors, i.e. muscle spindles. In the other cephalic muscles (at the exception of the extrinsic muscles of the eye) the muscle spindles are replaced by sensory nerve formations (of different morphologies and in different densities) and isolated nerve fibers expressing mechanproteins (especially PIEZO2) related to proprioception. This chapter examines the cephalic proprioceptors corresponding to the territories of the trigeminal, facial, glossopharyngeal and hypoglossal nerves.

Keywords: proprioception, muscle spindles, atypical proprioceptors, cephalic muscles, PIEZO2, mechanoproteins

1. Introduction

Proprioception is a quality of the somatosensory system that informs the central nervous system about the static and dynamics conditions of muscles and joints. This type of sensitivity has been studied in deep in the muscles depending on the spinal nerves and today the neurobiology of spinal proprioception is well known [1–4]. On the contrary, the neuroanatomy as well as the cellular and molecular bases of the proprioception in the cephalic muscles is not well known. Nevertheless, it is clear that cephalic muscles permanently develop fine adjustments of stretching and tone in facial movements, regulation of chewing force, oromotor reflex behaviors, verbal and nonverbal facial communication, swallowing, coughing, vomiting or breathing [5–7].

The skeletal muscles contain an intrinsic mechanosensory system, the proprioceptive system, which provides unconscious and conscious information to the central nervous system. The proprioceptive inputs originate in specialized sensory organs (proprioceptors) present in muscles (muscle spindles [8, 9]), tendons (Golgi's tendon organs [10]), joint capsules (Ruffini-like sensory corpuscles, Pacinian corpuscles and free nerve endings [11]), and presumably also the skin but their physiological properties suggest they are not the alternative to muscle spindles [2, 12–14]. The information encoded by the propioceptors gives rise to unconscious

IntechOpen

and conscious sensations, necessary for most basic motor functions [15]. For those interested in a recent review and in detail on both types of proprioceptors, we refer to the Banks [8] and Macefield and Knellwolf [16].

Some decades ago, Baumel [17] suggested that proprioceptive impulses from facial muscles are conveyed to the central nervous system via different branches of trigeminal nerve throughout multiple communications with the branches of the facial nerve. Actually, it is accepted that the proprioception of all cephalic muscles depends on the trigeminal nerve [6, 18].

Therefore, the first unresolved issue in cephalic proprioception is whether all cranial nerves that innervate striated muscles also collect their proprioceptive innervation. According to Lazarov [18] the proprioceptive innervation of all cephalic muscles depends exclusively on the trigeminal nerve. In other words: the sensory ganglia of cranial nerves lack of primary sensory neurons and the proprioceptors of the cephalic muscles are supplied by neurons from the trigeminal mesencephalic nucleus [19].

The second aspect pending clarification is: if the proprioception of the cephalic muscles depends exclusively, or mainly, on the trigeminal nerve, how do the fibers of this nerve reach the muscles of the territories of other nerves? This question can be answered because to extensive communications of the trigeminal nerve with other cranial nerves. The trigeminal nerve has numerous connections to the facial nerve [20–34] and the data collected from animal models indicate that the nerve fiber interchange is always from the trigeminal to the facial nerve and not on the contrary [35]. To serve facial proprioception additional connections between the facial and cervical spinal nerves exists [36, 37]. Apart from those communications no specific reference of communications between the trigeminal nerve with the glossopharyngeal, vagal and hypoglossal nerves were found. But presumably the trigeminal proprioceptive fibers pass from the trigeminal nerve to them directly on the target organs themselves (tongue, pharynx, palate) or through their connections with the facial nerve [28, 31, 32].

And the third main question of cephalic proprioception regards the identification and characterization of proprioceptors in the cephalic muscles. The skeletal muscles innervated by spinal nerves contain neuromuscular spindles and Golgi tendon organs, in addition to other types of corpuscles with less functional entity [8–11]. However, only the cephalic muscles supplied by the mandibular branch of the trigeminal nerve, and *the platysma colli* muscle contain neuromuscular spindles [38–40]. Therefore, cephalic proprioceptors, if any, have to be represented by other sensory nerve formations other than neuromuscular spindles. Recent studies, using immunohistochemistry techniques associated with specific markers related to mechanization, have shown that facial muscles [34, 41, 42] and some pharyngeal muscles [43] have differentiated sensory structures that presumably replace proprioceptors. However, it cannot be ruled out that sensitive nerve fibers reaching the muscles (especially nociceptive ones) can function as mechanoreceptors-proprioceptors (see [44]).

2. Distribution of typical proprioceptors in cephalic muscles

Typical proprioceptors of human cephalic muscles are represented by neuromuscular spindles as most of them lack Golgi tendon organs since they lack true tendons.

Muscle spindles have been found in muscles innervated by the trigeminal nerve while in the territory of the other cranial nerves, with very rare exceptions, are absent [6]. Recently, Junquera [45] determined the relative density of muscle

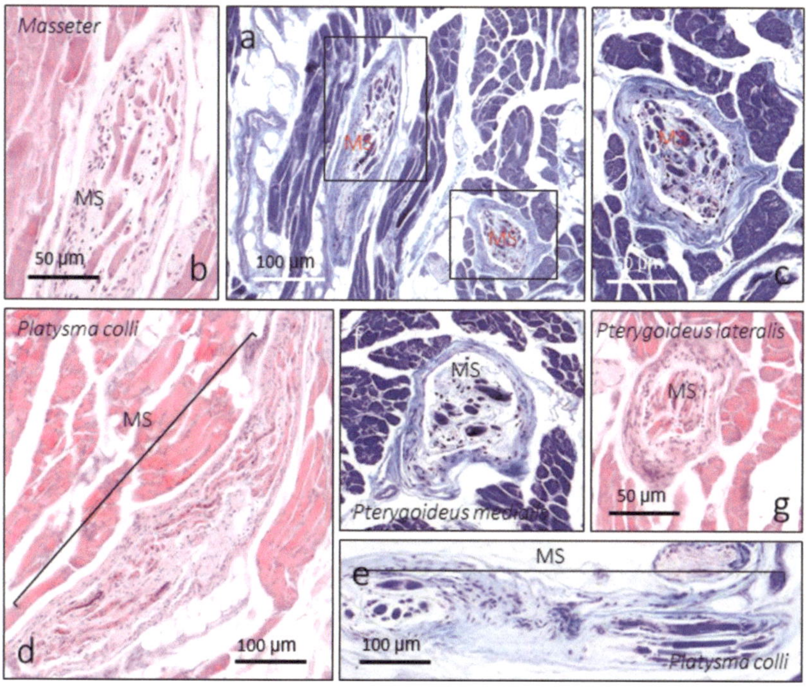

Figure 1.
Longitudinal and transversal sections of muscle spindles from different cephalic muscles. MS: muscle spinde.

spindles in human jaw muscles (**Figure 1**; **Table 1**). The *M. temporalis*, *m. masseter*, *m. perygoideus medialis* and m. *pterygoideus lateralis* contained numerous muscle spindles whereas they were less abundant in the *digastricus* and *mylohyoideus* muscles [45, 46]. The absence [45, 47] or presence [48] of muscle spindles in the *tensor veli palatini* muscle, also innervated by the trigeminal nerve, has been reported. It should be noted that atypical proprioceptors were also found in these muscles (**Table 1**; see below).

In muscles where the density of muscle spindles is higher, they consist of thick capsule, a shallow intracapsular space filled with variable number of intrafusal muscle fibers (ranging from 4 to 12). In muscles where the density of neuromuscular spindles was low, in general, the size of the spindles was smaller, had fewer intrafusal fibers and the capsule was less developed [45].

In the territory of the facial nerve one muscle spindle was found in the muscle *orbicularis oculi* in one pediatric specimen [49] whereas abundant muscle spindles have been found in the *platysma colli* [40]. Junquera [45] in her doctoral dissertation also observed typical muscle spindles in the *plastysma colly* more numerous in the cervical segment of the muscle than in the suprahyoid one.

3. Atypical putative proprioceptors cephalic muscles

3.1 Criteria to characterize atypical proprioceptors

The identification of putative sensory receptors in the cephalic muscles that may serve as proprioceptors was based on the following criteria: independence of the nerve trajectory, be placed in close relation to muscle fibers, show a morphologically differentiated aspect, and display immunoreactivity for any putative mechanoprotein [34].

Muscle	MS	Type I	Type II	Type III	INS*
M. temporalis	14	6	8	6	Yes
M. masseter	23	3	3	6	Yes
M. pterygoideus lateralis	18	3	14	7	Yes
M. pterygoideus medialis	21	5	10	3	Yes
Venter anterior m. digastricus	2	3	1	1	Yes
M. mylohyoideus	1	3	1	1	Yes
M. tensor veli palatini	0	2	2	1	Yes
M. corrugator supercilii + M. depresor supercilii		1	3	7	Yes
M. orbicularis oculii					
pars palpebralis	0	3	11	9	Yes
pars orbitalis	0	1	7	9	Yes
M. orbicularis oris					
pars marginalis	0	5	19	12	Yes
pars labialis	0	7	13	7	Yes
M. zygomaticus maior	0	1	4	4	Yes
M. zygomaticus minor	0	1	2	0	Yes
M. buccinator	0	19	28	10	Yes
M. depressor labii inferioris + mentalis	0	0	8	2	
M. levator labii superioris	0	1	1	3	Yes
Platysma colli**	12/8	11/7	4/7	6/8	Yes/ Yes
M. genioglossus	1	16	28	10	Yes
M. palatoglossus	0	0	5	3	Yes
M. uvulae	0	0	7	3	Yes
M. constrictor pharyngis superior	0	0	6	14	Yes
M. constrictor pharyngis inferior	0	0	5	9	Yes

**Isolated nerve fibers displaying immunoreactivity for any of the mechanoproteins investigated.*
***facial/cervical segments.*

Table 1.
Distribution and density of muscle spindles (MS), atypical proprioceptors (types I to III) and isolated nerve fibers (INF) in muscles supplied by the trigeminal nerve (green), facial nerve (blue), hypoglossal nerve (white) and glossopharyngeal nerve (brown).

In agreement with the above premises, capsulated and non-capsulated corpuscle-like structures of variable size and shape containing numerous axon profiles complexly arranged, have been identified. Given the morphologic heterogeneity of the corpuscle-like structures that fulfill the preestablished criteria we attempt to classify them into three types: type I, capsulated by a thin capsule, the glial cells variably arranged and showing different morphologies; type II, partially capsulated (the capsule being continuous with the perimysium), with variable morphology and in most of the cases the direction of the long axis was parallel to the one of muscular fibers; type III, non-capsulated and both the axon and Schwann-like cells are variably arranged (**Figure 2**).

On the other hand, it is now well established that at the basis of mechanosensitivity are mechanically-gated ion channels [50]. At present acid-sensing ion

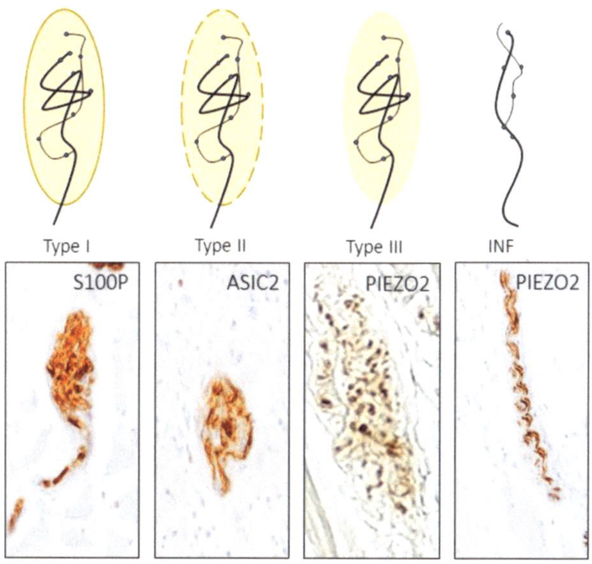

Figure 2.
Types of putative proprioceptors in human cephalic muscles. INF: Isolated nerve fibers.

channel 2 (ASIC2) and Piezo2 have been detected in muscle spindles and are strong candidates to initiate the mechanotransduction in proprioceptors [50–56]. Also, the putative mechanoprotein transient-receptor potential vanilloid 4 (TRPV4) was detected in proprioceptors of the facial and pharyngeal muscles [42, 43].

3.2 Distribution in the territory of the facial nerve

No typical muscle spindles have been found in the human facial muscles [42, 57–61] with the exception of on the facial part of the muscle *platysma colli* [40, 45]. Conversely, they contain numerous atypical proprioceptors (**Table 1**) the type II of being the predominating and the greater density being observed in the *buccinator* and *orbicularis oris* muscles.

3.3 Distribution in the territory of the glossopharyngeal nerve

Most research have not found typical muscle spindles in the muscles innervated by the glossopharyngeal nerve although they are present in the human *palatoglossus* muscle [48].

Regarding the pharyngeal muscles, typical muscle spindles were never found with the exception of the *constrictor pharyngis inferior* of the crab-eating monkey (*Macaca irus*) [62]. Nevertheless, human pharyngeal muscles are richly innervated. In particular, the *constrictor pharyngis superior* and muscle *constrictor pharyngis inferior* (innervated by branches of the pharyngeal plexus, derived from the glossopharyngeal and vagal nerves, and a small contribution of facial nerve; [63]) contain type II and III putative proprioceptors and isolated nerve fibers that display immunoreactivity for mechanoproteins (**Table 1**) [43].

3.4 Distribution in the territory of the hypoglossal nerve

As far as we know no muscles spindles have been reported in togue muscles. Junquera [45] observed one muscle spindle in the genioglossus muscle as well as numerous putative proprioceptors (**Table 1**).

Therefore, as a whole, the cephalic muscles have proprioceptive inervation, although only the muscles innervated by the trigeminal nerve and the *platysma colli* muscle innervated by the facial nerve contain neuromuscular spindles. The cephalic proprioceptors may be is involved in the coordination of facial movements and non-verbal communication, in language, swallowing and some other reflexes [64–66].

Author details

Juan L. Cobo[1,2*], Sonsoles Junquera[3], José Martín-Cruces[1], Antonio Solé-Magdalena[1], Olivia García-Suárez[1] and Teresa Cobo[2,4]

1 Departamento de Morfología y Biología Celular, Grupo SINPOS, Universidad de Oviedo, Oviedo, Spain

2 Instituto Asturiano de Odontología, Oviedo, Spain

3 Servicio de Radiología, Complejo Hospitalario Universitario de Santiago de Compostela, Santiago de Compostela, Spain

4 Departamento de Cirugía y Especialidades Médico-Quirúrgicas, Universidad de Oviedo, Oviedo, Spain

*Address all correspondence to: juancobodiaz@gmail.com

References

[1] Dijkerman HC, de Haan EH. Somatosensory processes subserving perception and action. Behav Brain Sci. 2007; 30: 189-201.

[2] Proske U, Gandevia SC. The proprioceptive senses: their roles in signaling body shape, body position and movement, and muscle force. Physiol Rev. 2012; 92: 1651-1697.

[3] Butler AA, Héroux ME, Gandevia SC. Body ownership and a new proprioceptive role for muscle spindles. Acta Physiol. (Oxf) 2016; 220: 19-27.

[4] Blecher R, Heinemann-Yerushalmi L, Assaraf E, Konstantin N, Chapman JR, Cope TC, Bewick GS, Banks RW, Zelzer E. New functions for the proprioceptive system in skeletal biology. Philos Trans R Soc Lond B Biol Sci. 2018; 373 (1759). pii: 20170327.

[5] Miller AJ. Oral and pharyngeal reflexes in the mammalian nervous system: their diverse range in complexity and the pivotal role of the tongue. Crit Rev Oral Biol Med. 2002; 13:409-425.

[6] Cattaneo L, Pavesi G. The facial motor system. Neurosci Biobehav Rev. 2014; 38:135-159.

[7] Frayne E, Coulson S, Adams R, Croxson G, Waddington G. Proprioceptive ability at the lips and jaw measured using the same psychophysical discrimination task. Exp Brain Res. 2016; 234:1679-1687.

[8] Banks RW. The innervation of the muscle spindle: a personal history. J Anat. 2015; 227:115-135.

[9] Bewick GS, Banks RW. Mechanotransduction in the muscle spindle. Pflugers Arch. 2015; 467:175-190.

[10] Ackermann PW, Salo P, Hart DA. Tendon Innervation. Adv Exp Med Biol. 2016; 920:35-51.

[11] Heppelmann B. Anatomy and histology of joint innervation. J Peripher Nerv Syst. 1997; 2: 5-16.

[12] Collins DF, Refshauge KM, Todd G, Gandevia SC. Cutaneous receptors contribute to kinesthesia at the index finger, elbow, and knee. J Neurophysiol. 2005; 94: 1699-1706.

[13] Macefield VG. Physiological characteristics of low-threshold mechanoreceptors in joints, muscle and skin in human subjects. Clin Exp Pharmacol Physiol. 2005; 32:135-144.

[14] Andreatta RD, Barlow SM. Somatosensory gating is dependent on the rate of force recruitment in the human orofacial system. J Speech Lang Hear Res 2009; 52:1566-1578.

[15] Akay T, Tourtellotte WG, Arber S, Jessell TM. Degradation of mouse locomotor pattern in the absence of proprioceptive sensory feedback. PNAS USA. 2014; 111:16877-16882.

[16] Macefield VG, Knllwolf TP. Functional properties of human muscle spindles. J Neurophysiol. 2018; 120: 452-467.

[17] Baumel J. Trigeminal-facial nerve communications. Arch Otolaryngol 1974; 99: 34-44.

[18] Lazarov NE. Neurobiology of orofacial proprioception. Brain Res Rew 2007; 56: 362-383.

[19] Yoshida A, Moritani M, Nagase Y, Bae YC. Projection and synaptic connectivity of trigeminal mesencephalic nucleus neurons controlling jaw reflexes. J Oral Sci. 2017; 59:177-182.

[20] Bendella H, Spacca B, Rink S, Stoffels HJ, Nakamura M, Scaal M, Heinen H, Guntinas-Lichius O, Goldbrunner R, Grosheva M, Angelov DN. Anastomotic patterns of the facial parotid plexus (PP): A human cadaver study. Ann Anat. 2017; 213:52-61.

[21] Shimada K, Moriyama H, Ikeda M, Tomita H, Shigihara S, Gasser RF. 1994. Peripheral communication of the facial nerve at the angle of the mouth. Eur Arch Otorhinolaryngol. S:110-112.

[22] Hwang K, Han JY, Battuvshin D, Kim DJ, Chung IH. Communication of infraorbital nerve and facial nerve: Anatomic and histologic study. J Craniofac Surg 2004; 15:88-91.

[23] Hwang K, Hwang JH, Cho HJ, Kim DJ, Chung IH. Horizontal branch of the supraorbital nerve and temporal branch of the facial nerve. J Craniofac Surg. 2005; 16: 647-649.

[24] Hwang K, Jin S, Park JH, Kim DJ, Chung IH. Relation of mental nerve with mandibular branch of the facial nerve. J Craniofac Surg 2007. 18:165-168.

[25] Hwang K, Yang SC, Song JS. Communications Between the Trigeminal Nerve and the Facial Nerve in the Face: A Systematic Review. J Craniofac Surg. 2015; 26: 1643-1646.

[26] Tohma A, Mine K, Tamatsu Y, Shimada, K. Communication between the buccal nerve (V) and facial nerve (VII) in the human face. Ann Anat 2004;186: 173-178.

[27] Hu KS, Kwak J, Koh KS, Abe S, Fontaine C, Kim HJ. Topographic distribution area of the infraorbital nerve. Surg Radiol Anat. 2007; 29:383-388.

[28] Diamond M, Wartmann CT, Tubbs RS, Shoja MM, Cohen-Gadol AA, Loukas M. Peripheral facial nerve communications and their clinical implications. Clin Anat. 2011; 24:10-8.

[29] Odobescu A, Williams HB, Gilardino MS. Description of a communication between the facial and zygomaticotemporal nerves. J Plast Reconstr Aesthet Surg. 2012; 65:1188-1192.

[30] Yang HM, Won SY, Kim HJ, Hu KS. Sihler staining study of anastomosis between the facial and trigeminal nerves in the ocular area and its clinical implications. Muscle & Nerve 2013; 48:545-550.

[31] Shoja MM, Oyesiku NM, Griessenauer CJ, Radcliff V, Loukas M, Chern JJ, Benninger B, Rozzelle CJ, Shokouhi G, Tubbs RS. Anastomoses between lower cranial and upper cervical nerves: a comprehensive review with potential significance during skull base and neck operations, part I: trigeminal, facial, and vestibulocochlear nerves. Clin Anat. 2014; 27:118-130.

[32] Shoja MM, Oyesiku NM, Shokouhi G, Griessenauer CJ, Chern JJ, Rizk EB, Loukas M, Miller JH, Tubbs RS. A comprehensive review with potential significance during skull base and neck operations, Part II: glossopharyngeal, vagus, accessory, and hypoglossal nerves and cervical spinal nerves 1-4. Clin Anat. 2014b; 27: 131-144.

[33] Tansatit T, Phanchart P, Chinnawong D, Apinuntrum P, Phetudom T, Sahraoui YM. A Cadaveric Study of the Communication Patterns Between the Buccal Trunks of the Facial Nerve and the Infraorbital Nerve in the Midface. J Craniofac Surg. 2016; 27:214-218.

[34] Cobo JL, Solé-Magdalena A, Junquera S, Cobo T, Vega JA, Cobo J. The proprioception in the muscles supplied by the facial nerve. En: *Selected Topics in Facial Nerve Disorders*, I Al-Zwaini y M

Jalal Husseis (Eds). Intech, London, 2019; pp. 1-14.

[35] Bowden RE, Mahran ZY. Experimental and histological studies of the extrapetrous portion of the facial nerve and its communications with the trigeminal nerve in the rabbit. J Anat. 1960; 94: 375-386.

[36] Yang HM, Kim HJ, Hu KS. Anatomic and histological study of great auricular nerve and its clinical implication. J Plast Reconstr Aesthet Surg. 2015; 68:230-236.

[37] Brennan PA, Elhamshary AS, Alam P, Anand R, Ammar M. Anastomosis between the transverse cervical nerve and marginal mandibular nerve: how often does it occur? Br J Oral Maxillofac Surg. 2017; 55:293-5.

[38] Scutter SD, Türker KS. The role of the muscle spindles in human masseter. Hum Mov Sci. 2001; 20: 489-497.

[39] Osterlund C, Liu JX, Thornell LE, Eriksson PO. Muscle spindle composition and distribution in human young masseter and biceps brachii muscles reveal early growth and maturation. Anat Rec (Hoboken) 2011; 294:683-693.

[40] May A, Bramke S, Funk RHW, May CA. The human platysma contains numerous muscle spindles. J Anat. 2018; 232: 146-51.

[41] Cobo JL. Neuroanatomía de la cara. Anatomía topográfica, quirúrgica, radiológica y microscópica de la inervación de la cara. Doctoral Thesis. Oviedo: Universidad de Oviedo; 2016.

[42] Cobo JL, Abbate F, de Vicente JC, Cobo J, Vega JA. Searching for proprioceptors in human facial muscles. Neurosci Lett. 2017; 640:1-5.

[43] de Carlos F, Cobo J, Macías E, Feito J, Cobo T, Calavia MG, García-Suárez O, Vega JA. The sensory innervation of the human pharynx: searching for mechanoreceptors. Anat Rec (Hoboken). 2013; 296:1735-1746.

[44] Ackerley R, Watkins RH. Microneurography as a tool to study the function of individual C-fiber afferents in humans: responses from nociceptors, thermoreceptors, and mechanoreceptors. J Neurophysiol. 2018; 120: 2834-2846.

[45] Junquera S. Propioceptores cefálicos: identificación basada en la expresión de PIEZO2. Doctoral Thesis. Oviedo: Universidad de Oviedo; 2020.

[46] Saverino D, De Santanna A, Simone R, Cervioni S, Cattrysse E, Testa M. Observational study on the occurrence of muscle spindles in human digastric and mylohyoideus muscles. Biomed Res Int. 2014; 2014: 294263.

[47] Kuehn DP, Templeton PJ, Maynard JA. Muscle spindles in the velopharyngeal musculature of humans. J Speech Hear Res. 1990; 33: 488-93.

[48] Liss JM. Muscle spindles in the human levator veli palatini and palatoglossus muscles. J Speech Hear Res. 1990; 33:736-746.

[49] Nelson CC, Blaivas M. Orbicularis oculi muscle in children. Histologic and histochemical characteristics. Invest Ophthalmol Vis Sci. 1991; 32:646-54.

[50] Kröger S. Propioception 2.0: novel functions for muscle spindles. Curr Opin Neurol. 2018; 31:592-598.

[51] Simon A, Shenton F, Hunter I, Banks RW, Bewick GS. Amiloride-sensitive channels are a major contributor to mechanotransduction in mammalian muscle spindles, J Physiol. 2010; 588: 171-85.

[52] Chen CC, Wong CW. Neurosensory mechanotransduction through

acid-sensing ion channels. J Cell Mol Med. 2013; 17:337-349.

[53] Gautam M, Benson CJ. Acid-sensing ion channels (ASICs) in mouse skeletal muscle afferents are heteromers composed of ASIC1a, ASIC2, and ASIC3 subunits. FASEB J. 2013; 27:793-802.

[54] Woo SH, Lukacs V, de Nooij JC, Zaytseva D, Criddle CR, Francisco A, Jessell TM, Wilkinson KA, Patapoutian. Piezo2 is the principal mechano transduction channel for proprioception. Nat Neurosci. 2015; 18:1756-1762.

[55] Florez-Paz D, Bali KK, Kuner R, Gomis A. A critical role for Piezo2 channels in the mechanotransduction of mouse proprioceptive neurons. Sci Rep. 2016; 6:25923.

[56] Anderson EO, Schneider ER, Bagriantsev SN. Piezo2 in Cutaneous and Proprioceptive Mechano transduction in Vertebrates. Curr Top Membr. 2017; 79: 197-217.

[57] Stål P, Eriksson PO, Eriksson A, Thornell LE. Enzyme-histochemical differences in fibre-type between the human major and minor zygomatic and the first dorsal interosseus muscles, Arch Oral Biol. 1987; 32: 833-841.

[58] Stål P, Eriksson PO, Eriksson A, Thornell LE. Enzyme-histochemical and morphological characteristics of muscle fibre types in the human buccinator and orbicularis oris, Arch Oral Biol. 1990; 35: 449-458.

[59] Kamen G, De Luca CJ. Firing rate interactions among human orbicularis oris motor units. Int J Neurosci. 1992; 64: 167-75.

[60] Happak W, Burggasser G, Liu J, Gruber H, Freilinger G. Anatomy and histology of the mimic muscles and the supplying facial nerve, Eur Arch Otorhinolaryngol. 1994; 1994: S85-S86.

[61] Goodmurphy CW, Ovalle WK. Morphological study of two human facial muscles: orbicularis oculi and corrugator supercilii. Clin Ant. 1999; 12: 1-11.

[62] Sengupta BN, Sengupta S Muscle spindles in the inferior constrictor pharyngis muscle of the crab-eating monkey (Macaca irus). Acta Anat (Basel) 1978; 100:132-135.

[63] Shimokawa T, Yi SQ, Izumi A, Ru F, Akita K, Sato T, Tanaka S. An anatomical study of the levator veli palatini and superior constrictor with special reference to their nerve supply. Surg Radiol Anat, 2004; 26:100-105.

[64] Wild B, Erb M, Eyb M, Bartels M, Grodd W. Why are smiles contagious? An fMRI study of the interaction between perception of facial affect and facial movements. Psychiatry Res. 2003; 123: 17-36.

[65] van der Bilt A. Assessment of mastication with implications for oral rehabilitation: a review. J Oral Rehabil. 2011; 38, 754-780.

[66] Schötz S, Frid J, Löfqvist A 2013. Development of speech motor control: lip movement variability. J Acoust Soc Am. 2013; 133: 4210-4217.

Chapter 6

Proprioception Impairment and Treatment Approaches in Pediatrics

Kamatchi Kaviraja

Abstract

In children problems like trauma and injuries are quite obvious. Other problems related to sensory system dysfunction are identified at the later stages of the child due to lack of awareness of the sensory integration problems which is not obvious. Some children have behavioral problems and some are poor at the school which is related to each other finally cause trouble to perform their daily routine. Early identification and intervention play a major role in improving the ability and development of the proprioceptive senses. Hence this chapter will introduce the new aspect of proprioception sense and its dysfunction. It would enhance you to identify the problems and understand the challenges that the child come across due to increase or decrease in proprioceptive input. We will be able to help them to overcome these challenges and frame a treatment strategy and help them to lead a successful life.

Keywords: proprioceptive dysfunction, sensory integration, gravitational insecurity, postural insecurity, modulation

1. Introduction

It is important to learn about the 7 senses. We all are familiar with the five senses and the other 2 senses are the Vestibular sense of balance, movement and Proprioception body position sense. By integrating, or combining all the information we get from our senses, we can 'make sense' of the world around us and successfully move through and interact in our world [1].

Proprioception is a continuous loop of feedback between sensory receptors throughout your body and your nervous system. Proprioception, also called kinesthesia, is the body's ability to sense its location, movements, and actions. Children who are clumsy, uncoordinated, and sensory seeking are often experiencing proprioceptive dysfunction. The following are common signs of proprioceptive dysfunction:

- Sensory Seeking (pushes, writes too hard, plays rough, bangs or shakes feet while sitting, chews, bites, and likes tight clothes)
- Poor Motor Planning/Control and Body Awareness (difficulty going up and down stairs, bumps into people and objects frequently, difficulty riding a bike)
- Poor Postural Control (slumps, unable to stand on one foot, needs to rest head on desk while working)

These children often self-regulate by engaging in behaviors that provide proprioceptive input such as toe walking, crashing, running or flapping. Heavy work or tasks that involve heavy resistance and input to the muscles and joints is essential to regulating proprioception [2].

Children that have difficulty sensing or processing proprioception often try to self-regulate by engaging in activities and/or behaviors that provide intense or frequent proprioceptive input [3].

Sensory Processing/Sensory Integration is when our different sensory systems work together to process different sensations from our body and/or environment. So, we are able to identify and give meaning to the different sensations we experience to accomplish daily activities and move in a coordinated manner. **Sensory Processing Disorder** is the inability to receive and efficiently use sensory information. Difficulty in processing sensory information interferes in our daily activities and impacts our functional ability to perform different tasks [4].

Benefits of sensory integration therapy may include: [5]

- Modulation of sensory systems
- Self-regulations
- Improved function in school, home and community
- Improved independence with activities of daily living (ADL)
- Maximized functional ability to perform daily and recreational activities
- Enhanced motor planning ability
- Active involvement and exploration of environments
- Efficient organization of sensory information

Assessment and intervention should explicitly focus on links among self-regulation, social participation, skills and perceived competence to address parents' expectations [6].

Interventions are proposed that relate to children's participation in contexts in which they live, learn, and play as well as the support of parents in the occupation of parenting [7].

Parents usually understand their child better than anyone else. They play an important role during the intervention and throughout their life time. I hope this chapter would enhance then to understand and give a clear picture of the child who have difficulty with proprioception dysfunction. The concept of this chapter comes from a body of work developed by A. Jean Ayres, PhD, OTR. This theory has been further developed and refined by the research of Dr. Ayres, and other occupational and physical therapists.

2. Proprioception impairment and treatment approaches in pediatrics

2.1 Proprioception

Proprioception refers to the sense of relative position and movement of the limbs and body.The mechanoreceptors embedded in the joint muscle, tendon and skin provides the proprioceptive information [8].

For good control of the muscle and voluntary movement proprioception is required. In pediatric conditions like CP, autism, Downs syndrome and development disorders it is always associated with proprioceptive deficits and hence the movement control is affected [9].

Proprioception provides ability to move. if there is any deficit in proprioception, our body movements would become slower, difficult and require more effort to perform a movement.

Processing of proprioception occurs:

CNS	**The spinal level**	The proprioception detects changes in the length and tension of the mucle and provides a stream of information to the cerebellum.
	Cerebellum	This information integrates with vestibular information. ↓ This contributes to postural control and a sense of gravity.
	Somatosensory Cortex	The proprioception information is integrated with the tactile system through dorsal column medial lemniscal pathway.

Proprioception always integrates with tactile and vestibular system [10].

Sherrington has the first studies on these topics in 1906, he has been established that muscle spindles are a major source of proprioceptive feedback to the central nervous system and appear to mediate the conscious perception of movement and limb position for the proprioceptive information [11].

The development steps:

1 Month	The newborn will interpret some of his body sensations and respond with built in reflex movement respond to • Gravity and movement • Muscle and joint • sensation • Sound • Smell & taste • Touch
2 & 3 Months	• Eyes and neck motor functions • Grasping • Raising up
4 to 6 Months	• Arms & hands movements • Airplane position • Recognizes the moving and loved to be moved
6 to 8 Months	• Locomotion • Spatial perception • Finger and eyes • Motor planning • Babbling
9 to 1 Months	• Play • Standing up • Words
2 Years	• Localization of touch • Moving • Mapping the body • Climbing • Self-hound
3rd to 7th Year	By 5 years the child becomes mature sensorimotor being.

The important age of development of sensory integration after birth is from 1st month to 7 years. Sensory stimulation, motor activity and exposure to environment during the early childhood have a great influence on the neurons and major role in the development of sensory and motor processes [12, 13].

2.2 Role of sensation

Sensations play a vital role in giving sensory input and information to the CNS. Every sensation is a type of information from the nervous system to produce a response and control the body and mind to the given information with a good amount of sensation or the sensory input is necessary for the development of the nervous system (**Figures 1–5**) [2, 14].

INTEROCEPTORS - Sensation which gives information about the inside of the body
↓
Visceral Sense

3. The process of sensory integration

Proprioception sense which is provoked by getting the information from contraction and stretching of the muscles, pulling and compression of the joints between bones.

The sensation from one's own body which occurs during movement both during static and dynamic and which always sends information to the brain about position is the proprioceptive sense [8].

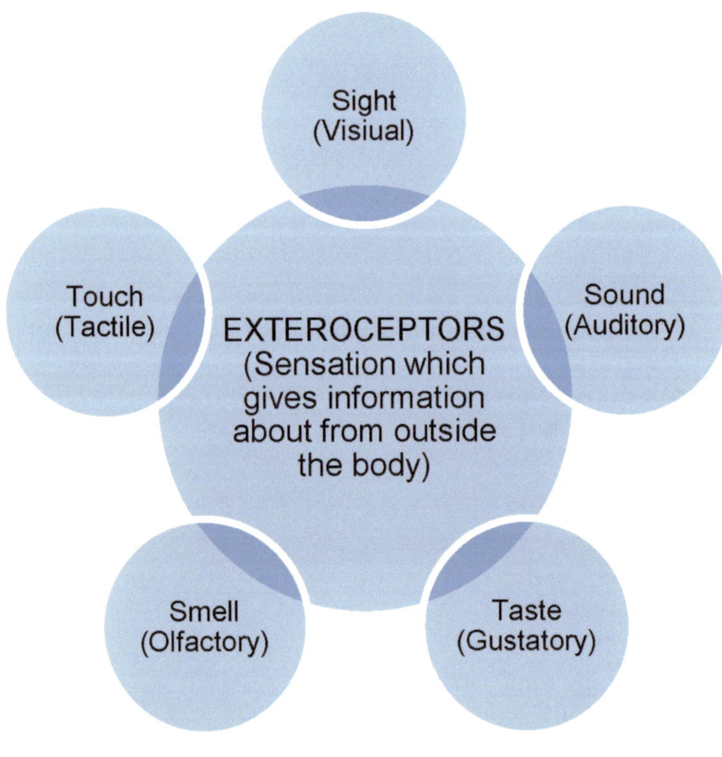

Figure 1.
Exteroceptors.

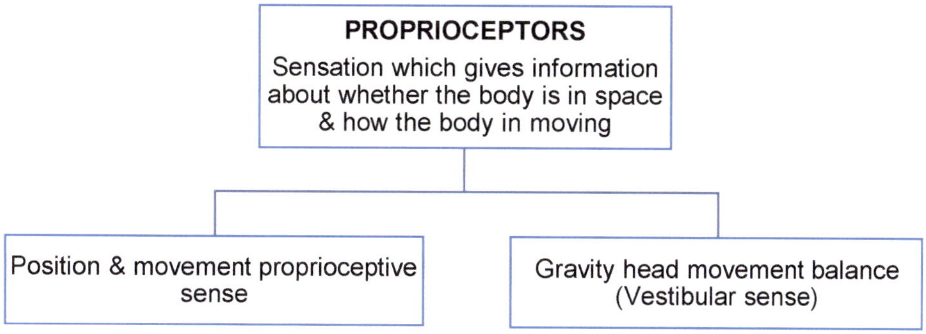

Figure 2.
Proprioceptors.

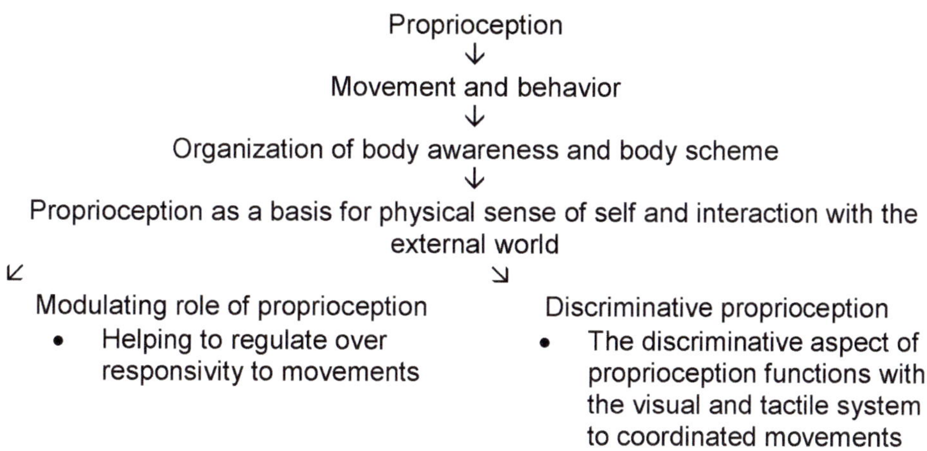

Figure 3.
Functions of propriocetion.

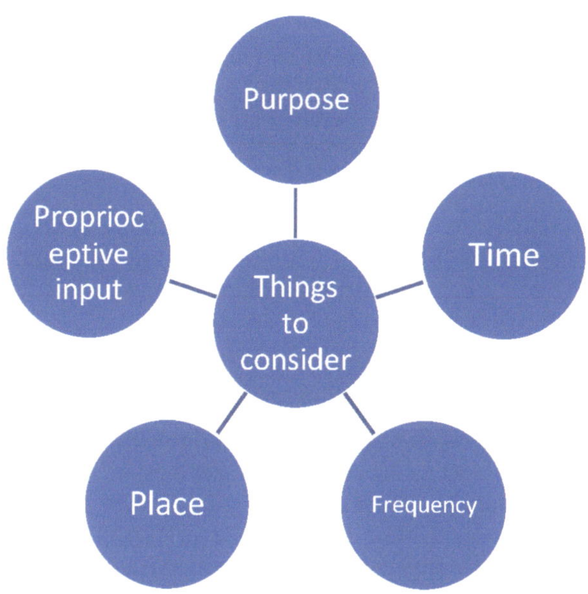

Figure 4.
Things to be consider.

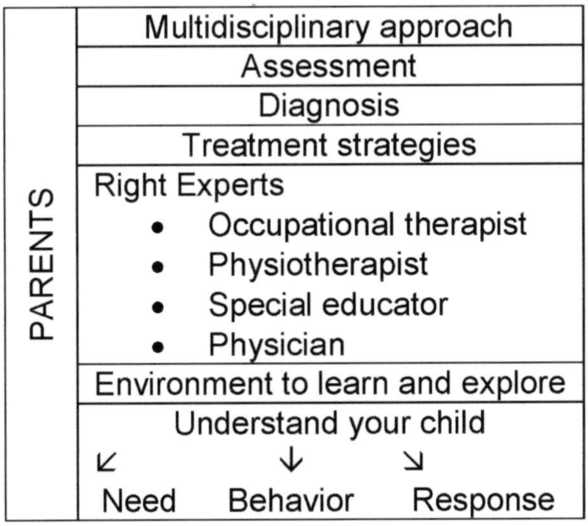

Figure 5.
For parents.

Proprioception is the unconscious awareness of the body in space or at resting position. The somatosensory system plays a major role in the sensory integration.

- The motor execution depends on somatosensory feedback.
- Important for body balance and praxis
- Important for reticular activating system.
- Important for the development of muscle tone.

Proprioception helps us to move. When proprioception is affected the child finds it difficult to perform the normal movement. The child finds it difficult to walk, clumsier, slower and must give more effort to perform the normal movement [15].

Any deficit or problem in the proprioception will lead to sensory integration dysfunction.

Causes,

- Hereditary factors
- Developmental disabilities
- Chemical factors

Due to lack of awareness about the neurological deficit, the parents will not be able to identify the problems or the difficulties faced by the children. They will not be able to analyze the learning and the behavioral problems are due to the sensory issue.

Sometimes lack of opportunities to play and explore and sometimes lack of interaction with the environment can also lead to these types of sensory seeking or sensory deficit in children.

It is always the responsibilities of the parents to give the child the required "sensory diet". Sensory diet is the required amount of sensory input that the child can gain from interacting with the surrounding and peer groups through playing etc. and explore new ideas and techniques to play and enjoy.

3.1 Functions

The modulating influence of proprioception over the senses appears to occur at the level of cerebellum, thalamus and somatosensory cortex [16].

4. Proprioceptive dysfunction

- Sensory seeking
- Poor motor planning/Control and body awareness
- Poor posture control
- Balance affected
- Uncoordinated movement
- Clumsiness.

The deficit associated with proprioceptive system.
The under reactivity to typical sensory stimuli:

- Hypo responsivity to proprioceptive input
- Usually exhibit hypo responsivity to touch
- Always seeks proprioceptive input to regulate or maintain state of arousal

Both over responsivity and under responsivity. Both extremes may occur in the same children [2].

- Gravitational in severity, Vestibular proprioceptive disorders
- Proprioceptive sensitivity.

4.1 Sensory diet

"The daily total of sensorimotor experience needed by a person to adaptively interact with the environment".

Sensory diet is for the self-regulation. The importance of the proprioceptive system in to give a person with information on how far to reach, how much pressure, where we are in space, to learn about body schema. It involves movement, compression and stretching at a joint.

Ayres conceptualized SPD as a disorder of body scheme in which children misperceive their immediate space and their surrounding space. She believed that therapy based on a "sensory integration approach" would normalize the spatial perceptions from multiple sensory systems and contribute to successful participation in daily life activities [17, 18].

Several authors have reported on the motor control difficulties related to poor proprioceptive processing among children with ASD, including decreased postural control and motor planning, overreliance on proprioception, difficulty matching proprioception with vision during reach, decreased organization of space, and poor motor anticipation [19, 20].

5. How to give proprioceptive input

Proprioceptive input can be given in two different ways one is calming and other is excitatory.

Calming activities is usually given to child who is over aroused.

Excitatory proprioception inputs are given to under arouse and the excitatory input should increase arousal state of the child.

It is important to assess the child's current state of arousal state of the child.

It is important to assess the child's current state of arousal.

Proprioception always woks along with the vestibular system which has a great influence and effect on the child behavior of the child.

Heavy work activities activate proprioceptive receptors

1. A big ball pit, bean bag, rolling up with blankets and pillows, jumping on bed

2. Trampoline jumps gives a great proprioceptive input

3. Wheel barrow walking

4. Pushing and pulling activities

5. Tight hugs, deep pressure

Figure 6.
Staircase walking with a ball. Ask the kids to climb up & down the stairs holding the ball for about 10 minutes 5 times followed by 1 minute rest 2 sets a day for 2 weeks.

6. Hand push, pull, compression

7. Wall pushes, star jumps

8. Squeezing a ball

9. Staircase walking with a ball *(***Figure 6***)*

10. Climbingwall bars, ropes

11. Throwing and catching weighted balls

12. Obstacle walking and crawling

13. Balance board activities *(***Figure 7***)*

14. Passing the ball *(***Figure 8***)*

15. Tug of war

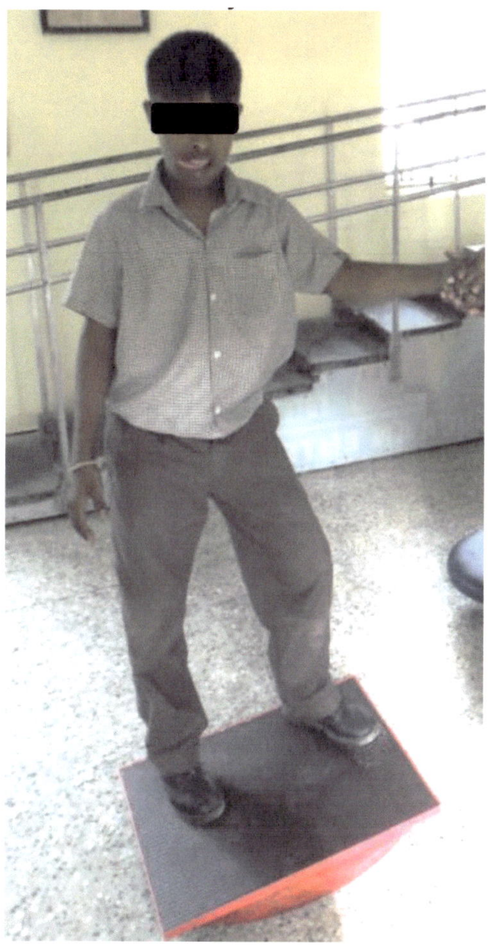

Figure 7.
Balance board activities. Ask the kid to step on the balance board 8–10 minutes first with support and then without support then to reaching activity twice a day 2–3 weeks.

Figure 8.
Passing the ball. Ask the kid to pass the ball by kicking 10–15 minutes twice a day for 2–3 weeks.

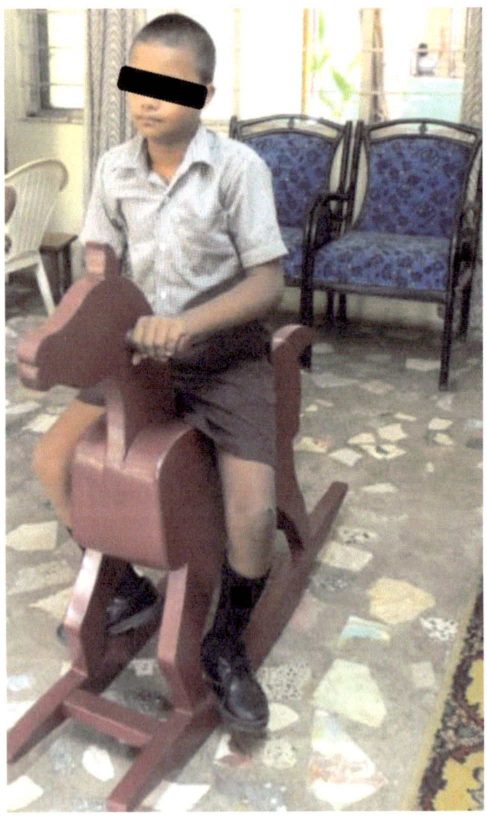

Figure 9.
Horse ridding. Ask the kid to sit comfortably and start rocking 10–15 minutes twice daily for 3 weeks.

6. Oral activities

1. Chewing

2. Blowing balloons

3. Blowing bubbles

4. Blowing small ball along a table

5. Oral massages

First can give full body movements.

- Frog leaps
- Horse riding (**Figure 9**)
- Crab walking
- Gorilla jumping
- Somersaulting
- Yoga

Continue with heavy workout

- Bear walks
- Cheetah runs
- Elephant stomp

The animal theme is a wonderful creative outlet and we can encourage the child to make animal sounds and gestures.

- Scooter board (**Figure 10**)
- Crawling on all fours
- Crushing activities
- Swiss ball activities (**Figure 11**)

The points to consider for giving proprioceptive input

- The purpose is to decide whether the activity to be given should stimulate an under responsive or sensory seeking.
- Secondly to identify the trigger points and suitable time to engage the child. We should identify when the child will get distracted.
- We should observe the child and analyses when she/he reach the calm alert state. This will help us to identify how long and how often activities should be given.
- Activities which can be done with parents at home can be taught to parents and the activities can be scheduled. Some activities like wall pushes, squeeze object, jumping etc.

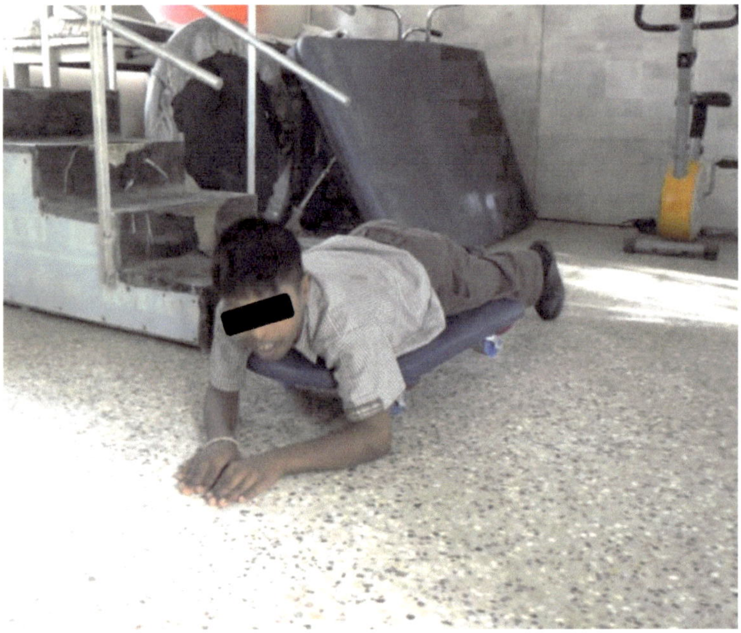

Figure 10.
Scooter board. Ask the kid to go for prone lying and using the hand pushing backwards to propel forward. 10–15 minutes twice daily 3–4 weeks.

Figure 11.
Swiss ball activities. Ask the kid to sit on the swiss ball & then bouncing followed by reaching activities 8–10 minutes twice daily 2–3 weeks.

- Therapy based exercise.
- Using a SI unit with various textures and different king of swings and mirrors to provide a visual feedback and the child can receive a maximum amount of proprioception input from the environment or the therapy room where it is well prepared for the child to receive the inputs.

The parents should try to analyses the behavior and the adaptive response that the child reacts or any changes which takes place before and after the therapy.

More often the sensory issues are mistaken for behavioral issue. The children always struggle with proprioceptive difficulties either hyperactive or decreased arousal level to perform their daily living activities.

First and foremost, the patient should analyses and whether the child avoid the proprioception inputs or seeing for proprioceptive input.

The parents should analyze and understand. If the child avoids the proprioceptive inputs

- Always the child will be lethargic
- Always tries to sit in a place
- Condition like autism the child will not have eye contact or social interaction
- The child will avoid physical activities
- Lack of coordination
- Inability to perform sports activities, climbing ladder and ropes
- Develops poor body posture

If the child seeks proprioceptive input

- The child always jumps and runs around
- Always on the move
- Aggressiveness
- Biting, kicking, hitting, pushing
- Always wants to chew and bite objects

A prolonged and multidimensional care is needed for the children [15].

There should always be an interaction between teacher, parent and therapist.

Sensory activities like deep breathing exercise with vital support. Sensory supports can be given with weighted blanked [21].

Assessment plays a major role in the development of the child integration. It is important to be aware of the assessment and evaluation required for the correct diagnosis for the child.

Therapist plays a major role to diagnose and to give the sensory integration Intervention.

- Identification of sensory integrative deficit
- Documentation of the level of function of the child
- Appraisal / reappraisal changes based on the outcomes
- Perform an informal assessment then with formal assessment the sensory integration and praxis tests "gold standard" for evaluating sensory integration and praxis functions.

SIPT is a standardized assessment tool with normative data for age limit 4–0 to 8–11 of age.

Some of the clinical outcome measures

1. OTA- Watertown clinical assessment which helps us to observe and examine sensory modulation and sensory discrimination.

2. Sensory profile (Dunn, 1994, 1999) Screens for SI dysfunction by assessing sensory responsiveness.

3. Sensory Integration Inventory (Rev. ed) (Reisman and Hansches, 1992) this screens for SI dysfunction.

7. Conclusions

Several studies have found evidence that children with sensory processing disorder can gain more insight into their development by assessing and providing an exact intervention at their right age and time. I hope this chapter would help out the parents to analyze the difficulties the children go through and guide them in the right path.

Acknowledgements

First of all, I praise the Almighty, who is above all for his glorious presence in me and for his blessings to complete this chapter.

I would like to convey my sincere thanks to Chancellor **Dr.A.C.SHANMUGAM, B.A., B.L., President Er.A.C.S.ARUNKUMAR, B.Tech, M.B.A, Secretary Thiru. A.Ravikumar M.B.A and Vice chancellor prof.Dr.S.Geetha Lakshmi** for giving me an opportunity to access the splendid resources in our university **Dr.M.G.R Educational & Research Institute University.**

I would like to express my sincere gratitude to the principal **Dr.C.V.SENTHIL NATHAN M.P.T(Geriatrics), PGDDR, M.I.A.P, MHCPC, MISCP, for his constant support and guidance. Dr.M.G.R Educational & Research Institute University.**

Finally, I would like to express my heartfelt thanks to my parents for their blessings and my beloved husband **Dr.N.KAVIRAJA B.P.T, DBDT, M.I.A.P, MRCI,MD(ACU)** and My son **K.JISHNURAJ** for their support &friends who have given many moments to cherish and treasure.

Conflict of interest

Nil.

Author details

Kamatchi Kaviraja
Faculty of Physiotherapy, Dr. M.G.R Educational and Research Institute, Velappanchavadi, Chennai, Tamil Nadu, India

*Address all correspondence to: kamatchi.physio@drmgrdu.ac.in

References

[1] Importance of Sensory Integration [Internet]. Vol. 501, Pathways. 2013. p. 2445. Available from: https://pathways.org/watch/importance-of-sensory-integration/#:~:text=Puttingtogether information from all,and interact in our world.

[2] Blanche EI, Reinoso G, Chang MC, Bodison S. Proprioceptive processing difficulties among children with autism spectrum disorders and developmental disabilities. American Journal of Occupational Therapy. 2012 Sep 1;66 (5):621–4. DOI: 10.5014/ajot.2012.004234

[3] Hansen A. What is Proprioception and Why is it Important? [Internet]. sonoran sun pediatric therapy. 2017. Available from: https://sonoransunpediatrictherapy.com/2017/11/16/what-is-proprioception-and-why-is-it-important/#:~:text=To put it simply%2C proprioception, attend and focus%2C and speech.

[4] What is Sensory Integration and Why is it Important in a Child's Development? [Internet]. Arizona Orthopedic Physical Therapy. 2013. Available from: https://azopt.net/sensory-integration/

[5] Sensory Integration Therapy [Internet]. American Academy of Pediatrics. 2012. Available from: https://www.healthychildren.org/English/health-issues/conditions/developmental-disabilities/Pages/Sensory-Integration-Therapy.aspx

[6] Cohn ES, Kramer J, Schub JA, May-Benson T. Parents' explanatory models and hopes for outcomes of occupational therapy using a sensory integration approach. American Journal of Occupational Therapy. 2014 Jul 1;68(4):454–62. DOI: 10.5014/ajot.2014.010843

[7] Cohn E, Miller LJ, Tickle-Degnen L. Parental hopes for therapy outcomes: Children with sensory modulation disorders. American Journal of Occupational Therapy. 2000 Jan 1;54 (1):36–43. DOI: 10.5014/ajot.54.1.36

[8] Buderath P, Gärtner K, Frings M, Christiansen H, Schoch B, Konczak J, Gizewski ER, Hebebrand J, Timmann D. Postural and gait performance in children with attention deficit/hyperactivity disorder. Gait & posture. 2009 Feb 1;29(2):249–54. DOI: 10.1016/j.gaitpost.2008.08.016

[9] Holst-Wolf JM, Yeh I, Konczak J. Development of proprioceptive acuity in typically developing children: normative data on forearm position sense. Frontiers in human neuroscience. 2016 Aug 29;10:436. DOI: 10.3389/fnhum.2016.00436

[10] Roley SS, Blanche EI, Schaaf RC. Understanding the nature of sensory integration with diverse populations. Pro-Ed; 2001.

[11] Sherrington C. The Integrative Action of the Nervous System. The Journal of Nervous and Mental Disease. 1907 Dec 1;34(12):801–2.

[12] Ayres AJ, Robbins J. Sensory integration and the child: Understanding hidden sensory challenges. Western Psychological Services; 2005.

[13] Tugay N, Tugay BU. Proprioception : The Forgotten Sixth Sense. Proprioception after Arthroplast. 2015;(June)

[14] Proprioceptive [Internet]. Middle Town Centre for Autism - Sensory Processing Resources. Available from: https://sensory-processing.middletownautism.com/sensory-strategies/strategies-according-to-sense/proprioceptive/#1

[15] Montgomery P, Richter E. Effect of sensory integrative therapy on the neuromotor development of retarded children. Physical Therapy. 1977 Jul 1;57 (7):799–806. DOI: 10.1093/ptj/57.7.799

[16] Should I Discipline My Child's Sensory Seeking Behavior? [Internet]. Brain Balance. Available from: https://blog.brainbalancecenters.com/should-i-discipline-my-childs-sensory-seeking-behavior

[17] Ayres AJ. Development of the body scheme in children. The American journal of occupational therapy: official publication of the American Occupational Therapy Association. 1961 May 1;15:99–102.

[18] Ayres AJ. Sensorimotor foundations of academic ability. Perceptual and learning disabilities in children. 1975;2: 301–58.

[19] Weimer AK, Schatz AM, Lincoln A, Ballantyne AO, Trauner DA. "Motor" impairment in Asperger syndrome: evidence for a deficit in proprioception. Journal of Developmental & Behavioral Pediatrics. 2001 Apr 1;22(2):92–101. DOI: 10.1097/00004703-200104000-00002

[20] Schmitz C, Martineau J, Barthélémy C, Assaiante C. Motor control and children with autism: deficit of anticipatory function?. Neuroscience letters. 2003 Sep 4;348(1):17–20. DOI: 10.1016/S0304-3940(03)00644-X

[21] Christy E. Yee O. Sensory Diet Chart. 1995;1–2

Chapter 7

Proprioceptive Perception: An Emergence of the Interaction of Body and Language

Alejandra Vasquez-Rosati and Carmen Cordero-Homad

Abstract

This chapter provides a systemic perspective of human behavior, which reformulates the concept of effective behavior and cognition that derive from the classical vision of neuroscience and psychology based on the Cartesian reductionist functionalist paradigm. This systemic perspective, which is based on the theory of autopoiesis, proposes that the act of perceiving proprioception is decisive in the capacity of the human being to differentiate himself from an external space within which he is situated; a phenomenon that we will denominate "proprioceptive perception". This complex phenomenon of dynamic character emerges from the relationship between the domains of the body and language in the individual's relationship with their environment. Furthermore, from this systemic perspective, we will present the emotional states as cognitive states necessary for the conservation of the individual's living identity and the close relationship they have with the sensorimotor patterns and proprioceptive perception. This chapter answers the question of how proprioceptive perception affects the human being's experience of being different from others and from the environment in which they find themselves, having the possibility of being aware of themselves and of the world they perceive - in a present - within the environment in which they find themselves. And it explains how this phenomenon modulates its modes of emotion in congruence with what occurs in its present.

Keywords: autopoiesis, three-dimensionality of behavior, emotions, sensorimotor correlation, posture, sensorimotor pattern, emotional states, proprioception, perception

1. Introduction

The proprioceptors are sensory receptors that refer to the qualities of movement in the postural dynamics and displacements of the body in space; sense that for a human being is determinant in the perception of himself in a here and now. We call this phenomenon proprioceptive perception, which modulates the emotional states of the individual given the circumstances of the present.

Proprioceptive perception is a complex phenomenon of dynamic character that results from the modulation of phenomena of different orders such as physiological, relational and interpretative. In this chapter we will address the questions of how, in the epigenesis of a human being, proprioceptive perception affects his capacity to become aware of his corporeal existence within his contextual situation with others and with the environment in which they exist, moment by moment; and how the proprioceptive experience modulates -in a present- his emotional states in relation to the immediate interactive context maintaining physiological states of the organism congruent to the present circumstances, conserving in adaptation his life in the space in which it exists.

To answer these questions, we will first situate ourselves from a dynamic systemic perspective that reformulates the concept of effective behavior and cognition that derive from the classical vision of neuroscience and psychology based on the Cartesian reductionist functionalist paradigm. This last one understands cognitive phenomena from a representational perspective, where cognition is conceived as an information processing that results in a faithful representation of an external world that operates independently of the organism that perceives it. This has kept science in search of an understanding of the principles and laws of an objective external world, which explains why in the study of perception, there is a prevalence over the exteroceptive senses of an individual (vision, hearing, touch, smell and taste), ignoring the incidence in the sensory integration of the proprioceptive and interoceptive senses. In second place, we will approach how in the origin of the first living organisms are constituted the generative sensorimotor mechanisms of the movements of the effective behaviors that reveal the knowledge of the living beings, to know that it is source and origin of the way of knowing proper of the *Homo sapiens*: the reflection. Next, we will explain how the synchronization between attentional reflexive movements and corporal movements gives origin to the proprioceptive perception that makes possible the differentiation of the external and internal space of the individual. And following with the phenomena of the emotions, we will explain the characteristics of the ways of moving of a human being and how the proprioceptive perception influences the modulation of these in relation to the conservation of the well-being of the organism in its structural coupling with the environment. It explains the concept of emotional plasticity and the type of practices that restore it, showing evidence of its effectiveness.

2. The paradigmatic shift

We will begin with the concepts that articulate the reflexive logic of this study that approaches the phenomena of cognition from an evolutionary systemic look that comes from the Theory of Autopoiesis based on the “Theory of the Biology of Knowledge” of Humberto Maturana and Francisco Varela [1], which brings about a radical paradigmatic shift that is produced with the evidence that the internal operation of living beings -in their environment- is of a circular and recursive nature. This implies that the cognitive processes are referred to the changes of the internal states, and not to the changes of the external environment within which it is observed, showing how the world they live in, in a present, is the result of an epigenic process and not of a processing of information captured from the environment. This explains the phylogenetic and ontogenetic origin of the cognitive processes of human beings that - in their relationship with others - give rise to the domain of language, which makes possible their capacity to reflect, and with it, to perceive themselves differentiating proprioceptively from others and from the world of objects that they learn to perceive in their culture.

Both authors define that “knowledge is effective action, that is, operational effectiveness in the domain of existence of the living being” [1], specifying that this domain is constituted, moment by moment, in the physiological operation of the living body in interaction with the environment in which it exists. They show how in organisms with motility this operation of a cognitive nature determines the changes of state of the individual in relation to the conservation of his living identity in a changing environment, and not to changes in the environment, revealing that effective behaviors do not respond to the perception of an image or representation of the state of an external world.

Therefore, we will address the explanation of how the first unicellular organisms of the planet autonomously maintain effective or adequate behaviors for their conservation within the environment from which they arise and with which they maintain a continuous interactional relationship, a fact that reveals their knowledge of how to live in a changing environment, thus showing how the cognitive capacities that, as we will see, result from the physiological operation of the organism in its interaction with its environment, are constituted in the evolution of the species.

This reflection originates with the study of living beings -including human beings- as dynamic autopoietic systems [1]. That is, as living systems that are self-generating, moment by moment, referring to the dynamic organization of molecular relations that constitutes them. This organization remains invariant in a flow of internal structural changes within a changing environment, with which it maintains a continuous interactive relationship of reciprocal nature, preserving the organization that defines its identity as a species.

Thus, in the operation of this organization, a network of interactive relationships between molecular components that produce the components that constitute the - metabolic network - is constituted, moment by moment, maintaining an operation of a circular and recursive nature that generates autonomy, which determines, moment by moment, the appropriate internal states for the conservation of the molecular organization that constitutes them as living beings within the changing environment in which they exist [2]. In this way, the dynamics of this molecular system constitute a physiological operation of cognitive nature, that maintains the orders of the interactional relations between the molecules, making possible the existence of a living unity that differentiates from the environment that it exists in a permanent reciprocal interaction with it.

Therefore, all living bodies, with or without nervous system, are autonomous self-referential beings, with the ability to determine the appropriate behaviors for their living conservation. So, the changes of states of the organism that trigger internal or external disturbances to themselves, are specified by their internal autopoietic operation and not by the changes of the external environment; environment with which it maintains a continuous of recurrent and recursive interactive relations of reciprocal character, that is, bidirectional.

Thus, the reciprocal character of the mutual interactions between the living body/niche generates in the epigenesis of the individuals a congruence or correspondence between the structures of both spaces constituting a structural coupling, in which both spaces -delimited by the edge of the organism- modulate with each other without there being control of one over the other, since each domain specifies the structural changes triggered by the disturbances produced in their mutual interactions, moment by moment.

In this way, in this study, the living body/niche dynamic interactional system is considered as the unit of study of an individual's behavior; two subsystems that constitute the operational domains of behavior: a) the body domain: constituted in the operation of the physiological dynamics that constitute it, and b) the relational domain: generative of the interactional dynamics that are generated in its operation within the environment in which the individual as a whole exists, an interactional space in which behavior is observed.

"...the phenomena of the structural dynamics of a living system and the phenomena that occur in its interactions in the medium, are phenomena of different kind that occur in phenomenal domains that do not intersect, and cannot be expressed one in terms of the other". Thus, "[...] behavior of a living system is the interactional and relational dynamics through which a living system realizes its living as a particular kind of organism in its domain of existence [...] the structural dynamics of the living systems triggers structural changes in the medium, and at the same time the structural changes that take place in the medium as behavior takes place trigger structural changes in the living system. As living takes place in the continuous conservation of autopoiesis and adaptation by the living system through its behavior, the behavior of the living system operates as the guide in the conservation or loss of the living through the coupling of the structural dynamics of the living system and the medium." [3].

In this way "what we call behavior when observing changes in the states of the organism in its environment corresponds to the description of the movements of the organism in an environment that we point out." [1]. This means that this environment does not correspond to the world in which the individual lives.

The implication of evidencing the autonomy of living systems, brings a radical epistemological paradigmatic shift by modifying the conception of living beings, since an autonomous system means that it defines itself through mechanisms of self-organization. Therefore, this characterization of living beings modifies fundamental beliefs of the traditional Cartesian, representational and functionalist paradigm that conceives living systems as heteronomous systems, that is, that they are defined - in their conformation and behaviors - through external mechanisms of control (input–output), therefore their world is treated as if it were independent and represented [4]. In this way, classical science defines behavior as responses to external stimuli, being the environment the one that defines the course of structural changes of the living bodies, thus living bodies have no incidence in their evolutionary transformations, for which they would be heteronomous systems. This is exemplified in the following statement by H. Curtis and N. Barnes:

"The characteristics of the behavior of an organism -its sensitivity to particular stimuli and patterns of response to those stimuli- are the product of natural selection, just as much as the shape of the teeth or the feedback loop that regulates blood pressure. Therefore, natural selection is the force or active agent that determines the course of evolution of the identity of living beings, being these mechanical organisms lacking the autonomy to specify their behavior and structural changes in relation to the conservation of the molecular organization that defines their living identity, and actively specify the niche in which they carry out their living through their behavior, conserving themselves in adaptation in a structural coupling with the environment." [5].

The reformulation of the generative mechanisms of the effective behaviors leads in turn to a reinterpretation of the concept of cognition, which traditionally has been considered as an information processing in which the sensorial surfaces transduce the stimuli of the environment, sending the information to neuronal structures that process it, generating a representation of the state of the world in which the individual is, from which the system selects the effector motor patterns for the appropriate behaviors to the individual's situations, which result from phylogenetic and ontogenetic learning of an adaptive nature.

The autopoietic theory originates a reformulation of the generative mechanisms of effective or adequate behavior of living beings, as well as the evolutionary processes that give rise to the diversity of anatomo-physiological structures that define the identity of living species. The autonomy of living bodies makes them an active agent in the transformation of themselves, as well as of the environment in which they are found; a phenomenon that occurs in the epigenesis of living bodies. The evolution of the history of structural links between organisms and their environment generates reciprocal transformations that lead to the maintenance of a congruence in the structural changes of both, a fact that makes possible the conservation in adaptation of the organisms. Traditionally, this congruence has been interpreted as the result of effective behaviors in their adaptation to the environment, which result from responses to the stimuli of an external world that operates independently of the operation of the organism, considering them instructive and unidirectional interactions; thus, in the evolution of the organism, the adaptive behaviors would be determining the structure and identity of the species.

3. The origin of the autonomous movement and the lived world

The sequences of movements of an organism in coupling with the environment that are observed in its effective behaviors result from "a very specific correlation coordination between sensory surfaces and motor surfaces, ...sensory-motor correlations that originate from the first living beings through metabolic transformations proper to the cellular unit" [1], which in the recurrence and recursiveness of reciprocal interactions with the environment are constituted in sensory-motor learnings that specify ways of interacting in the regularities of changes in the environment, keeping invariant the molecular relations that define the molecular organization that preserves the living unity in adaptation.

By way of example, the feeding behavior of an amoeba about to ingest a protozoon is described by means of the extension of a pseudopod. Pseudopods are expansions or digitations of protoplasm associated to structural changes in the local physicochemical constitution of the cellular membrane. How does the global and unified movement of the animal occur in its structural coupling with an environment in which it is also structurally coupled to it? "The presence of the protozoon generates a concentration of substance in the environment that is capable of interacting with the amoeba's membrane, triggering changes in protoplasmic consistencies, resulting in the formation of a pseudopod. The pseudopod in turn produces changes in the position of the moving animal, thus modifying the number of molecules in the medium that interact with its membrane. This cycle is repeated, and the sequence of displacements of the amoeba, therefore, is produced through the maintenance of an internal correlation between the degree of modification of its membrane and those protoplasmic modifications that we see as pseudopods, a recurrent and invariant correlation is established between a disturbed or sensory area of the organism and an area capable of producing motor displacements, which maintains invariant a set of internal relations in the amoeba." [1].

We can see, on the one hand, how the continuous structural coupling of the organism with the environment generates the congruence between the structural changes of both, and on the other hand, how the movements of the organism generate correlations of specific structural changes between sensory and motor surfaces that establish interactive relations of reciprocity. These relationships are not instructive, they generate a continuous structural change in which the change of one is in relation to the change of the other, moment by moment. Thus, the changes in the motor surfaces generate, in turn, changes in the sensory surfaces, sensorimotor dynamics generating permanent movements that are observed in the proper behavior of an organism in its environment, a process that, as we see, does not consist of a process of capturing and processing information from an external world that operates independently of the organism's operation.

In this way, in the physiological operation of the organism, the modulation of processes of sensorimotor activity is generated, constituting a synchronic coordination of structural changes between local zones of the organism that modulate with each other, resulting in a distributed modulation mechanism from which a state of global activity emerges -of a temporary nature. This global state specifies a coherent and unified cognitive state that determines the behavior of the individual in his/her relationship with the environment; a mechanism in which local changes modulate the state of global activity and, vice versa global states modulate the activity of local areas, without the existence of an external or internal agent or force that controls them. Such mechanisms of sensorimotor coordination are constituted in the operation of every living being with or without nervous system -what varies among species are the types of sensorial and effector structures- of a centralized control or product of an external or internal agent that specifies the states of activity of sensorimotor patterns of the organism, as well as dispenses with the idea of a representation of an external world.

Returning to the behavior of the amoeba in its coupling with the environment in which the protozoon is found, they establish a structural coupling, which from an observer's perspective, the protozoon constitutes the prey as a result of a feeding behavior. This fact that a western human being perceives, observes and interprets from the distinctions of the world of his culture, has no relation with the intentional behavior of the amoeba, since by the way, the amoeba does not distinguish the protozoon nor has perspective of the changing environment, therefore it does not intentionally go towards it with the purpose of swallowing it and thus feeding itself. This anthropomorphic interpretation hides what occurs inside the animal in its structural coupling with the environment, that is, it hides the physiological operation of a cognitive nature of the organism that specifies its changes of states triggered by the disturbances of the environment, in relation to its previous states, a fact alluded to when characterizing said operation as circular and recursive.

In summary, from this systemic perspective, the sensation-movement state of a living body in its environment, in a present, results from the dynamic activity of the sensory-motor operations that gives rise to a global cognitive state - of a temporary character - that specifies the coordinated and synchronic dynamic movement sequences that constitute the coherence and uniqueness of an effective or adequate behavior for the conservation of the organism in adaptation in the structural coupling with the environment. This was illustrated with the case of the amoeba's behavior, showing the origins of the generative mechanisms of the effective behavior of living beings with motility, as well as the cognitive processes that result from their co-evolution with the environment in which they exist.

From what has been said before, we can distinguish that the environment in which an observer distinguishes the amoeba, does not correspond to the world that it lives from its sensorimotor dynamics that are referred to its internal operation, which on the one hand means that the environment that it knows is its interiority and on the other hand that these dynamics that specify the movements in their structural coupling with the environment are generative of the lived world, a world without perspective of its changing environment.

We could say that the living of a body arises spontaneously in a generative movement of its knowing, which Maturana expresses strictly by saying: "to live is to know and to know is to live" [1], which from our perspective alludes to a fundamental fact that reveals the mode of existence of every living being, namely, both life and knowledge arise in the same act.

Therefore, the autonomous movement of animals with motility is a key to the understanding of cognition and the phenomenon of perception in human beings, as we will see in the following section the knowledge of the body in its environment is the source and origin of the way of knowing of human beings: reflection. In this regard the biological origin of human knowledge is evidenced by unifying its nature: corporeal-spiritual constitutive of a living unit.

From the world of the biology of knowledge we reach the world of philosophy. The phenomenologist Maurice Merleau-Ponty from his exhaustive studies in the human experience, describes the phenomenology of perception, a study that begins with the conviction that "phenomenology is also a philosophy that re-situates the essences within existence and does not believe that man and the world can be understood only from its factuality" [6]. From his studies of the human perception and behavior, he establishes co-relations between the psychism and physiology that lead him to a reformulation of the classic vision of the body-object, saying: "The union of the soul and the body is not sealed by an arbitrary decree between two external terms; one, the object, the other, the subject. This union is consummated every moment in the movement of existence. It is existence that we find in the body when we approach it through a first way of access, that of philosophy." [6].

4. The co-evolution of living beings

Bearing in mind that the world that a living organism feels within its coupling with the environment, it constitutes a continuum of sensation-movement resulting from the cognitive states that emerge from the dynamics of activity of patterns of sensorimotor correlations. These patterns, which are constitutive of the learning process, result from recurrent and recursive movements that are constitutive of the individual's behavior, and determine his anatomo-physiological structure, which specifies his species and his way of knowing and living in his structural coupling with the environment.

In the -recurrent and recursive- structural couplings between living beings that co-exist in the same environment, it is constituted temporary reciprocal interactions between them, generating mutual learning that modify in congruence the anatomo-physiological structures of them. For this reason, in each temporary encounter between them the autonomous operating of the corporal structure of each one determines the specific movements of their behavior, recreating the structural couplings that occur in these temporary encounters. For example, this phenomenon occurs with symbiosis relations between species. This is the case with the structural correspondence between pollinating insects and the flowers of the plants they pollinate. The plant species Drakeae glyptodon, an orchid species whose structure takes a similar form to the female Thynnid wasp, and in its operation produces pheromones that attract the male wasp which is the only insect vector of its pollination. Thus, in the co-evolution of both species, they constitute a history of structural coupling that constitutes the structural changes that are conserved in their progeny.

This epigenic phenomenon when it occurs between individuals of the same species gives rise to the constitution of social systems. In the recurrent encounters of two or more organisms, specific action dynamics are synchronically triggered, generating a coordination of action between them resulting in a communication that specifies a particular way of interacting and relating, which defines a domain of possible actions between them. Thus, said systems are constituted in dynamics of networks of coordination of action between individuals, that give emergency to a collective of autonomous beings self-organized, which moves like a totality in congruence to the changes of the environment, inside which the individuals generate behaviors that of isolated form they could not acquire. This is the case of the flocks of Franklin gulls that migrate from Cape Horn to Canada, a flight in which individuals increase their speed of flight by 72%, compared to the speed of flight of an isolated individual, and no further in the case of human beings who, in their social way of life, learn in doing with another to incorporate the mastery of language that makes their capacity for reflection possible.

Thus, in social systems, the learnings that are generated in individuals in the coordination of action among them, constitute the sensorimotor patterns that are the ways of moving and relating that constitute the way of life of the collective, which is transgenerational preserved by maintaining a living knowledge that makes its existence possible within the environment.

This co-evolutionary phenomenon constitutes a communicative process that is not related to an exchange of information, but rather to interactions of a reciprocal nature that generate specific and recurrent structural changes that occur in their encounters; encounters in which the structural changes of the organisms in their reciprocal interactions are modulated - at each instant - generating a coordination of movements that configure a choreography that is repeated in their recurrent interaction within the social system in the environment in which they are found. "We will understand communication as the mutual triggering of coordinated behaviors among the members of a social system" [1].

5. Reflexive movement and proprioceptive perception

Following this second order cybernetic perspective [7], which recognizes that the architecture of the neural networks constituting the nervous system that is embedded in the body of the organism, maintains a circular operation that is to say with operational closure, therefore this operation is referred to the states of activity of the network, and not to an external world [8]. This network is self-organized by distributed mechanisms, in which the global states of activity of the network modulates the activity of local zones, and vice versa the activity of local zones modulates the global states of the network. In this way, there is no internal or external agent that controls its operation; on the other hand, this system modulates and is modulated by the physiological operation of the organism. Therefore, the condition of operational closure of the neuronal network would explain that the world perceived by the organism, including the human being, is a world that emerges from the internal operation of the organism in its structural coupling with others and the environment.

Considering this, we will explain the perception of the world lived by a human being and the learnings that originate the proprioceptive perception, from recognizing the type of structural links that occur between the hominid ancestors of *Homo sapiens*, generative of phylogenetic learnings that make possible the emergence of language and its capacity to reflect. These facts give origin to the particular way of life of a social system constituted by networks of coordinated action coordinations from the operational distinctions of the language domain, that is to say, generative action coordinations of the networks of consensual conversations. We are going to see how the first condition for the constitution of the phenomenon of perception in human beings is the origin of the observer. "The observer and the observed, then, emerge in the flow of structural changes that occur in the members of a community of observers when they coordinate their consensual actions through their recurrent structural interactions in the domain of operational coherences in which they carry out their connected practices of living." [9]. Thus, if language is constituted in the domain of the coordinations of the dynamics of action coordinations that occur in the joint action of individuals within the social system in which they carry out their living in coexistence.

The mathematician H. Von Foerster, in his original presentation of the notion of second-order cybernetics, who is a precursor of the same, starts by pointing out what he calls a theorem, alluding to the statement made by Maturana after explaining the origin and mode of operation of language, which he states in this way:

"I. - All that is said is said by an observer.

A theorem to which Foerster adds a corollary that affirms:

II.- All that is said is said to an observer".

Concluding that I and II connect two observers through language, with this connection, in turn, a new concept is established, namely that of society, the two observers constitute the fundamental nucleus of a society, thus the three concepts are connected in a triadic way, each one with the other. "In this interaction we cannot say who was first and who was last [...] a closed triad is formed" [7].

Thus, in this circularity, the operation of language within a social system constitutes observers, and observers in turn constitute the language in its operation in the domain of the coordination of the coordination of action among observers, a reflexive dynamic that is recurrent and recursive and generative in turn, of the domain in which it is constituted. This is a dynamic in which perceptual objects that constitute the world of culture in a society of human beings are configured. Thus, the operation of language domain constitutes an operational system constituted by the distinctions of perceptive objects that are associated among them, generating new distinctions that constitute the association network that generates them, a circular and recursive operation.

In the operation of language, the observer and the perceptual objects that constitute him/her are constituted, a circular dynamic that generates the world perceived by individuals, which does not correspond to an objective reality. Therefore, the description that an observer makes of the world of objects or phenomena that he/she perceives is the result of the flow of the experience of his/her consensual behavioral coordination with others. Therefore, these descriptions are not absolute truths, but descriptions agreed upon with others in the coexistence by such "everything said is said by an observer and for an observer" with whom they maintain a generative structural congruence of their coordination of actions in doing together in the coexistence.

"And since perceptual objects arise as behavioral configurations, the world of shared perceptual objects belongs to the sphere of operational concordances between organisms, which constitute them in the course of their coexistence as configurations of their behavioral concordances. In other words, if the perceptual objects remain configured by the behaviors of the organism, the world of perceptual objects that occurs in the coexistence of organisms, including the observer, can only arise from the coexistence as long as the organisms operate generating and conserving their mutual structural correspondence. That this is so is also apparent in everyday life, in which we know that the common world only arises in the community of living" [10].

How does the observer in language generate perceptual objects that are configured in the behavior of the individual? Language occurs in the flow of consensual coordinations of actions of organisms whose actions are coordinated because they have congruent dynamic structures that have emerged or are emerging through their recurrent interactions in a co-ontogenetic structural drift. Because of this, interactions in language are structural interactions that trigger in the organisms interacting contingent structural changes with the course of the coordinations of consensual actions in which they arise. As a result, even though the domain of language is not intercepted with the structural domain of the body of the interacting organisms, the structural changes of the interacting organisms in language are a function of what occurs in their language and vice versa [9].

In this way, the origin of language generates a new operational domain in the behavior of human beings, which generates reflexive operations. Thus, this domain, which is not intercepted with the corporeal domain (constitutive of the physiological operation), nor with the relational domain (constitutive of the reciprocal interactions that the organism maintains with others and its environment), is constituted as a domain that in its operation generates perturbations both in the state of the organism and in the interactive contexts of the organism. Therefore, it corresponds to a third operational domain, which participates in the modulation of human behavior and experience.

The operation of language generates associations, descriptions and interpretations that originate the beliefs of the world of culture that give meaning to the way of doing and relating to individuals, making them learn to incorporate the recurrent and recursive coordination of action. This operational domain, brings the intentional movements from the reflection, which entails the learning of the reflexive movement of orienting the focuses of attention towards the perceptive objects constitutive of the world that the individuals learn to see in the doing with others, within the culture in which they grow. This reflexive attentional movement brings the possibility of the human being to become aware of himself, of the others and of the environment in which they are, by differentiating himself proprioceptively from the objects that he perceives, which occurs by the ways of relating and interacting that are constituted in the way of life of the hominids.

We will now see what happens in the bodily domain of behavior with the structural couplings in the hominid lineages that give rise to the phylogenetic learnings that make the origin of *Homo sapiens* possible. In the lineages of hominids that give rise to the human being, their way of life was generated learning gave rise to the architecture a nervous system that is characterized by a significant increase in brain mass, which means an increase in interneurons that expands the possibilities of structural plasticity of individuals, and thus the ability to learn, which means greater behavioral plasticity. Today we know that the genomes between *Homo sapiens* and anthropoids are almost identical, and from neuroscience it is observed that the regions of the brain have equivalents in the brains of apes. An interesting difference is the development of the generative auditory capacity of phoneme learning that is related to the origin of language [11]. Therefore, the advances of science support with their data the assumption that the origin of the reflection capacity of human beings is related to their way of life.

Such learning, which modifies the anatomophysiological structure, occurs in a way of life in which the game generates continuous and recurrent coordinations of action, and in this way increases the capacity to manipulate and differentiate objects with which they interact with others. At the same time, they had daily physical encounters in which they groomed themselves, and caressed in sensual and prolonged interlacing of their bodies and continuous sexual games with prolonged physical contact. In this way these dynamics of action generate structural changes in proprioceptive sensorial surfaces that correlate with modifications in motor surfaces, constituting recurrent changes in the sensoriality of the qualities of the movements of dynamics of postural sequences and positions of the different parts of the body, which together with maintaining a frontal vision with the other in the coordination of movement that are accompanied by guttural sounds, establishing in dynamics of action in which they generate movements of joint generative attentions of a perspective of the movement of itself and the other, which entails the reflective learning to sustain a division of centers of attention in movements of visualization of the movement of the other and proprioceptive sensation of the movements that it coordinates with him. Thus, in these dynamics of structural links of character -recurrent and recursive- learning is produced, which are constitutive of sensorimotor patterns that result from the coordination of visual, proprioceptive, tactile and auditory sensations, which constitute the sensation of movement and space of oneself, which occurs simultaneously with the differentiation of others in space.

This pre-reflective process that is observed in the families of hominids is generating the learning that makes possible in the individuals the reflective movement of coordination of the attentional movements and the corporal movements in relation to one another, generative of a perspective that arises from an internal space delimited proprioceptively towards an external space when dividing its attention. "...in effect, their spatiality is not, like that of external objects or like that of spatial sensations, a spatiality of position, but a spatiality of situation. [...] The word here, as it applies to my body, does not designate a certain position with respect to another position or with respect to an external coordinate, but rather the installation of the first coordinates..." [6]. In this way, his corporeality is a spatial reference of his situation in his perspective of the world, moment by moment, which arises in interactions with others and the environment, being the place perceived proprioceptively in which he is and exists in a present.

This reflexive movement constitutive of an observer's perspective, which arises from learning to divide their focus of attention into movements of joint attention with another observer in the manipulation of their bodies and objects, is configuring perceptual objects that constitute the observer that emerges with them. In this way the reflexive movement that arises from motor couplings between individuals, is constitutive of the operational domain of language and of the structural congruence between the objects of perception and the living body in its structural coupling with the environment: "The observer's operation in language consists of a way of living in the recursion of behavioral coordination that arise in the community of living and that configure a world of perceptual objects. [...] The language and the operation of the observer, therefore, do not require or give rise to references to an external reality. The world of the observer's descriptions and explanations is a world of modes of coexistence that generate perceptual objects, in which the observer emerges as one of them when language emerges. Hence the generative and transforming power of the world that language and the explanations given in it have." [12].

6. Proprioceptive perception

This composite concept leads us first to consider how perception is understood from this systemic perspective that includes living beings as structurally determined systems. This means that everything that happens to the organism in the interaction with its environment is determined by its structure. Therefore, the interactions of the organism with its environment are not instructive [1, 12]. From this perspective, perception consists of "the configuration that the observer makes of perceptual objects by distinguishing operational cleavages in the behavior of the organism, by describing the interactions of the organism in the flow of its structural correspondence in the environment" [12].

For there to be a perceptive experience, it is the observer who emerges from language as a perceptual object, the one who configures a world of perceptual objects in the recursion of behavioral coordination that arise in the community of living [13].

On the other hand, we will define proprioception as one more sense like vision, smell, taste, hearing; it is the sense of the qualities of the body's movement and its situational disposition in space (that the same movements generate) (see **Table 1**). Proprioception is not in itself a form of perception that gives us the "perception of the body", it is not the image, nor the representation, nor the consciousness of the body as an object [14]. The proprioceptive sensation is produced, moment by moment, by the changes in the activity of the proprioceptors that generate the

Location	Proprioceptor	Quality of sensation
Muscle	Spindle afferents Ia & II	Length, speed, acceleration and deceleration Minimal over-contraction force.
	Golgi tendon organ	Dynamic changes of the contraction force
	Group III y IV	Chemosensitives. Information on metabolic changes and muscle damage/inflammation
Joint	Group I & II	Range, speed and position of the joint. Group I (dynamic and static, low threshold, slow adapting), Group II (dynamic, fast adapting)
	Group IV	Feedback on excessive stress on the joint. Sensitive to joint inflammation
Skin mechanoreceptors	5 types of receptors in the skin: two fast adapting and three slow adapting	Contact and texture of objects. The tension of the skin contributes to the sense of movement of the joint. More sensitive to dynamic than static stimulation

Table 1.
Proprioceptors and quality of sensation.

dynamics of the postural sequences of the movement of the individual in structural coupling with the environment. This dynamic phenomenon in which the relations of reciprocity between the changes of the sensorial surfaces and the effector surfaces of the movement, generate that the sensation modulates the movement and the movement in turn, modulates the sensation, a continuous flow of sensation-movement. This flow of sensation-movement is constituted in the operation of sensorimotor patterns that specify qualities of behavioral movement. In the reciprocal interactions of the individual in structural coupling with the environment, the cognitive states specified and that in turn specify, the changes in the dynamics of activity selectivity of the sensorimotor pattern networks that give rise to sensory integration (proprioceptive, visual and vestibular) [15] that define the dynamic body scheme, in a present.

The body schema is defined as an integrated set of sensorimotor processes that organize perception and action in a non-conscious and sub personal way [16]. The body schema is not phenomenologically available to the observer: "the body schema is not the perception of my body, it is not the image, the representation or even the consciousness of the body. Rather, it is precisely the style that organizes the functioning of the body in communion with its environment [17]. On the other hand, body image includes the immediate conscious perception of the body, including the conceptual construction about the body and the emotional attitude and feelings about the body, "being a complex phenomenon that contemplates at least three aspects: perceptual, cognitive and emotional" [17]. However, other definitions have been proposed for this construct: "cognitive representation of the body based on stored knowledge and sensory experience that underlies perceptual judgments" [18], "a representation of the body's shape" [19], "perception of the body's spatial dimension, its size, shape and relative configuration of its parts" [20].

What are we talking about when we talk about proprioceptive perception? Proprioceptive perception differs from the concepts of body schema and body image, since it is a reflexive phenomenon that constitutes an attentional movement of the observer towards the corporeal dimension of his behavior, in a here and now. Thus, proprioceptive perception makes present as object of perception the proprioceptive qualities resulting from the dynamics of postural movement and displacements of the individual in his structural coupling with the environment.

These qualities configure the perception of the dynamic corporeal space that is defined in a flow of synchronic coordination of movements of the different parts of the body that configure the coherent and unified global movement constitutive of the proprioceptive qualities that result from the sensorimotor operation of the individual in his structural coupling with the environment.

Thus, proprioceptive perception is the perceptual object of the observer configured with the qualities that make up the internal space that appears sensorially delimited from an external space within which it is situated, generating a perspective of the world of objects from which it differs proprioceptively, perceiving the place in which it exists, in zero time; that is, the living body that constitutes moment to moment, its existence as a living being in a structural coupling with the environment with the capacity to reflect and observe the world that it constitutes in doing with others within its culture.

"If corporeal space and outer space form a practical system, the latter being the background against which it can stand out, or the void before which the object can appear as an objective of our action, it is evidently in the action that the spatiality of the body is carried out, and the analysis of one's movement has to allow us to understand it better. We understand better, as soon as we consider the body in movement, how it inhabits space (and time, for that matter), because movement is not satisfied with passively supporting space and time, it actively assumes them, it takes them back in their original meaning that is erased in the banality of acquired situations." [6].

In this way, proprioceptive perception cannot be understood outside of perception-movement. Proprioceptive perception constitutes the reflexive and corporal movements of two dimensions of human behavior constitutive of disjointed operational domains: language and its corporeality. "Reciprocally, every perceptive habit is still a motor habit and here also the capture of a meaning is made by the body." [6].

So proprioception does not have a dual nature, as proposed by Gallagher [15], since its nature is biological and responds to physicochemical properties. Proprioception corresponds to the body domain; whose operations are the networks of physiological dynamics that constitute the mechanisms of the correlations of the sensory and motor surfaces. While reflection and movements of the focus correspond to the domain of language, whose operations are the networks of semantic distinctions with operational closure. Therefore, when proprioception is a perceptual object of the observer, both the body and language domains are operating simultaneously on proprioceptive perception. In addition, these disjointed domains modulate each other [21], and reflection and attentional movements can trigger changes in proprioception and in turn proprioception generates changes in the language domain, as we will see later.

Consequently, we say that the phenomenon of proprioception is different from the qualities of the perceptual object that the observer configures from his corporeal experience, which results from the modulation of the three operational domains that configure the coherence and uniqueness of his behavior: corporeal, relational and language, moment by moment. Thus, both the proprioception and the proprioceptive perception of the individual in their interactive contexts maintain a structural congruence between both phenomena in their continuous structural changes within their circumstances, thus constituting the effectiveness of their behaviors in relation to both their purposes and the conservation of their well-being.

For this we will first address how muscle physiology is involved in the modulation of body perception. The situational disposition of the individual (his posture and movements, in a present) correlates with a configuration of the afferent activity of the proprioceptors coming from the skin, the joints and the muscles that are projected towards the primary somatosensory cortex and the primary motor cortex, to then converge in higher order somatosensory regions [20]. The integration and comparison of proprioceptive activity with the activity of other sensory modalities (and the reflective capacity of the human being) triggers the perception of the size of the body parts, which is relative to the perception of other body parts, as well as to the environment in which the individual is coupled in a present. Thus, in situations where the activity of the nervous system presents a change in the relationships that are generative of its structure, as is the case of a vascular accident, epilepsy, anesthesia or migraine, the perception of size and shape of body parts will be modulated by this configuration, which is commonly understood as a perceptual "illusion" of the body. This phenomenon has also been observed by applying an external vibration in specific muscle regions [22]. Since the afferent activity of the muscle is modulating the sensation of the position of the limb, when performing such stimulation, it is possible to generate the "illusion" of the perception of the movement of the limb or the whole body in a desired virtual direction.

In these cases, the perception of the body is modified by unintentional factors on the part of the individual. However, the human being, through his reflective capacity, has the ability to direct his attention to the perception of his body and with it modulate the perception of the relative size and shape of his body parts. The evidence shows how paying attention to proprioceptive sensations (directing attention to movement during the execution of a task) generates a change in the sensitivity of the muscle spindle [23, 24], which would be modulating the perception of movement of the individual in its structural coupling with the environment. In this sense, training the proprioceptive perception we can modulate the muscular physiological activity, which as we will see, modulates in turn the sensorimotor correlations of the basic emotions.

7. The modulation of proprioceptive perception and emotions

In research on emotions, we find a diversity of explanations that involve descriptions of different mechanisms that affect the emotional states of a human being, which respond to different dimensions of the phenomenon: physiological, psychological, relational, behavioral, as well as cultural. Thus, in 1991, Plutchik in his book Emotions [25] indicates more than 57 definitions that arise from various authors in the field of physiology and psychology, such as W. James, S. Freud and B. Skinner, to mention a few. This fact shows the multiplicity of non-linear variables that characterize an emotion, so we can conclude that it is a complex phenomenon, which is naturally observed in the behavior of an individual, and that each person perceives in his experience.

Given this last point, we will understand "emotions as specific sequences of movement of an organism in structural coupling with the environment that an observer distinguishes". We approach emotional phenomena as the distinction of a specific configuration of a coherence in behavior. In this way we distinguish the phenomena that occur in the different operational domains of behavior: body, relation and language, and correspondingly we observe the correlations of the modes of movement, relationship and interpretation of an individual's experience.

These specific sequences of movement that constitute modes of movement define possible dynamics of action of the individual in his or her present, and with this the type of interactions that are generated in his or her relational contexts, as well as the distinctions of perceptual objects that originate his or her attentional movements in language, generating his or her interpretations.

In the human being two orders of emotional phenomena are observed that respond to the origin of sensorimotor learning, we find the basic emotions of phylogenetic origin, those -fear, rage, joy and sadness, on which the ontogenetic learning constitutive of the secondary or social emotions are interwoven [26], in the present study only the first ones are approached.

In the basic emotions, patterns of movements generate the activation of specific muscular synergies that are triggered from the autonomic nervous system, and therefore correspond to physiological and cognitive states of the organism. Damasio et al. [27], studied the activity of the central nervous system during the evocation of memories of the 4 basic emotions. In this they observed a specific activation pattern at cortical and subcortical level for each one of the emotions. Furthermore, they observed that the emotional states evoked activate the anterior pontine nucleus, which sends projections to the cerebellum and therefore, would possibly be involved in the activation of specific sensorimotor patterns and the quality of movement of each basic emotion. These findings show that each emotion has a physiological configuration of the nervous system and the motor system that is unique to each state, which correlates with a global cognitive operation that gives rise to the "knowing" of the organism in relation to its environment.

The specific movement sequence patterns we are talking about, correlate with specific sensorimotor patterns that come from a phylogenetic learning, that is, they are sensation-movement patterns that we can identify even in primitive unicellular organisms. Thus, the simple expansion and contraction movements of living bodies are indicative of the approach-avoidance behavioral pattern observable from a cell to the human being [28]. Therefore, from the sensory-motor operation of the organism in its structural coupling with the environment, emerges "the knowing" that is evidenced by the autonomy of the body to determine its effective or adequate behaviors to the maintenance of its living and social identity. That is to say, "knowing" emerges with the minimum living unit that moves and feels, feels and moves constituting the basic emotional movements that preserve the way of being of a living being within an environment that it does not know.

Therefore, these emotions that underlie every secondary emotion are related to the conservation of the individual's living identity, so that in continuous flow of the changes of state of the organism in its structural coupling with the environment, an emotional state of a cognitive nature can be identified, through the identification of the movements that generate the muscular synergies that are activated autonomously by the physiological operation of the organism. From here we speak of these emotions as a living knowledge that guides our actions in relation to preserving the essential, life.

These basic emotional movements correspond to fear, anger, joy and sadness, which are differentiated by a set of qualities of the sequence of their movements and the activation of muscular synergies [29]. A recent study by Shafir et al. [30], from the analysis of the movement of each one of these emotional states, identified those crucial motor elements that distinguish each emotion and that in turn, in their repetition are capable of evoking an emotional sensation. The results showed that each emotion is predicted by a single set of motor elements and that each motor element is a predictor of a single emotion, suggesting that the 4 emotions under study are discrete and have a biological substrate (see **Table 2**).

These motor patterns for each emotion delimit the possible movements of the individual, determining specific dynamics of action in its structural coupling with the environment, which in turn determine the individual's modes of relationship. Therefore, in the observation of an individual's mode of movement it is possible to characterize these modes of relationship from the flow of postural movement dynamics generated by each emotion. These dynamics are distinguished in the

Emotion	Quality of emotional movement
Rage / Anger	Advance with sudden, direct effort. Punching movements and leaning forward.
Fear	Locking up and condensing the body, as well as receding into space and retracting into the shape of the body.
Sadness	Passive weight sadness, sinking (letting the ribcage fall), head down, drooping shoulders and arm(s) to upper body, loss of muscle tone
Joy	Jumping and rhythmic movements. Lightness (light) and free flow. Movements that enlarge the body in a horizontal and vertical direction and upward movements in space.

Table 2.
Emotions and movement qualities (adapted from [30]).

experience from the proprioceptive perception, because the quality of the movements in which they generate the dynamics of activity of the muscular synergies, – speed and direction of the movement, force and muscular tone- are specific in each emotional disposition, for such in the lived experience patterns of emotional perceptions are evoked registered from proprioceptive perceptions that are correlated with the states of the evoked body.

Thus, the human being with his capacity of reflection, can recognize an emotional state in himself through the proprioceptive perception of the sensation-movement of the body of his emotion, in a present. From the study of the emotional experience, it was shown how the proprioceptive perception plays a central role in the identification of the sensations associated with global states of the body, giving emergence to the emotional experience [31]. The execution of specific body movements evokes emotional states related to those movements [32]. In turn, an emotional state modulates afferent muscular activity, modifying the patterns of sensation-movement. These observations confirm that the continuous modulation of the behavior and experience of a human being is constituted in a joint and disjointed operation of the three operational domains: body, relation and language.

In the study of Shafir et al. [30] they show that the repetition of a movement is capable of evoking an emotion, the attention is directed to the execution of that movement or sequences of movements, therefore, proprioceptive perception is active. In this way, if from the reflective movement of the attention, proprioceptive perception is intended in a present, the emotion is modulated in relation to the immediate environment and not to the flow of evocative associations of a past or future, generating a greater congruence in the structural coupling with the environment in which the living body exists, in a present.

The aforementioned is confirmed by the results of our studies about emotional plasticity in people who practice the cognitive body integration method (CBI), which correspond to a movement-based contemplative practice [33]. CBI practice is constituted from the model of the three-dimensionality of behavior to which we have made reference in this chapter. In the research we measured the autonomic response, through the pupil diameter, during the presentation of images with emotional content in a group of people who had experience in CBI practices and in a control group (CG). Our results showed that the CBI group presented shorter pupil recovery times than the CG group, showing a better emotional adaptation given the context of the individual, in a present [34].

The concept of emotional plasticity alludes to the natural loss of generative behavioral plasticity in the epigenesis of the individual, due to the history of structural links with others and the environment in which they are placed. This generates ways of moving, doing, and interpreting that are proper to the way of life of the family and culture in which the person lives, configuring in their behaviors modes of emotion that maintain a prevalence of a basic emotion, which over time restricts the domains of action of people, often reaching states of distress and loss of wellbeing within the current way of life. Thus, from the model we call "three-dimensionality of behavior", correlations between the three operational domains of behavior are distinguished, generating correlations between ways of doing, relating and interpreting of people; from which personalized practices are designed. These practices consist of exercises in which the movement of attention towards the body – in a recursive and frequently manner- is synchronized with dynamic recurrent and recursive movements that involve the master muscles of the muscular synergies of an emotion, with reflections of what occurs in the present. Thus, these practices are intended primarily to restore emotional plasticity in people, and generate learning to modulate their emotional states, from intentional attention to proprioceptive perception, which facilitates placing oneself in the space within the environment in which one exists, maintaining a state of presence in the here and now of the body, which gives an emotional autonomy that modulates the physiological states congruent with the present contingencies, maintaining well-being in the sense of coherence with the present situation and not only of joy or enjoyment.

8. Conclusions

The purposes of approaching the paradigm from which the reflexive logic of the explanations of our observations of the phenomena of human behavior and experience is generated are, on the one hand, to show how the explanatory models and their concepts configure the perceptual objects of the world that we perceive, in this case from the doing of science. And on the other hand, to show how the recognition of the autonomy and self-reliance of the body, which reveals the knowledge that results in the continuous structural coupling of the organism with its environment, gives us a look at how the harmonies or orders that are given in the co-evolutionary drift of living species are generated, which allows us to have new references to evaluate the incidences in the individual and collective well-being of the ways of doing and relating of people in the current way of life.

In relation to our study, we can conclude that, in the epigenesis of an individual, a structural congruence is generated -between proprioception as an operation of the body and the configuration of proprioceptive perception in the domain of language- generating a co-determination of both phenomena in the structural coupling of the individual with others and his immediate environment, in a present. This explains that proprioceptive perception is not a dual phenomenon, but emerges from the interaction of the three operational domains of behavior as a coherent and unified experience. Proprioceptive perception - as the perceptual object of the observer in language - modulates and in turn is modulated by the muscular physiology that from its structural changes specifies qualities of movement observed in individual behavior and that in its experience are configured as qualities of movement, volume, relative dispositions of parts of the body and relative to their situation in space.

Proprioceptive perception has great implications for the modulation of an individual's mode of emotion, which are defined by specific physiological states. This occurs because the dynamics of specific movements of each base emotion - which characterizes the way of moving -, are related to the conservation in adaptation of the individual within his changing environment, in a present, and not to the interpretations that he makes of his situation, which is the case of secondary or social emotions, those that do not present defined physiological and cognitive states and respond to cultural learning. Therefore, proprioceptive perception places the individual in a situation within his present circumstances, which occurs in conjunction with reflexive attentional movements of a generative character of an incorporation in the field of attention of proprioceptive perception. This attentional movement is correlated with changes in the motor surfaces that modulate their way of moving in congruence with the circumstances of the environment. This explains why contemplative practices that intend attentional reflexive movements together with body movements decrease the states of stress, which from our perspective is a generative physiological alteration of the secondary emotions that respond to the associative flow of language.

Consequently, assuming that the cognitive processes of both language and the body maintain an operational closure, we postulate that proprioceptive perception as a perceptual object is configured by spatial and movement qualities that correlate in the body domain with structural changes of the sensory and motor surfaces of the corporeal self in its interaction with the environment. Thus, the self in its history of structural coupling with its environment generates the sensorimotor learnings constitutive of the proprioceptive structure and networks of attentional selectivity that make possible the perception of a delimited internal space that originates its external space, which correspond to a space in which its existence is constituted and in which it exists, bringing to the hand the possibility of taking a perspective of itself, which occurs when differentiating proprioceptively from others and from the changing environment in which they exist. In other words, behavior and the experience of the lived world are co-determined in the interactive operation of the three operational domains of a disjointed character of behavior. And as we see the environment in which an observer distinguishes an individual, it does not correspond to his or her lived world.

From this proposal, new interesting topics are opened to deepen the understanding of these phenomena: the relations that are constituted between the reflexive movement of attention and body movements of the individual in relation to the configuration of the proprioceptive perception.

Author details

Alejandra Vasquez-Rosati[1,2]* and Carmen Cordero-Homad[2]

1 Body Phenomenology Laboratory, Chile, Brazil

2 Cognitive Body Integration Center, Santiago, Chile

*Address all correspondence to: alejandravasquezrosati@gmail.com

References

[1] Maturana H, Varela FJ. El Árbol Del Conocimiento: Las bases biológicas del entendimiento humano. Editorial Universitaria; 1984.

[2] Maturana HR, Varela FJ. De Máquinas Y Seres Vivos. Autopoiesis: La organización de lo vivo. [Internet]. 5th ed. Chile: Editorial Universitaria; 1998. Available from: http://www.ncbi.nlm.nih.gov/entrez/query.fcgi?cmd=Retrieve&db=PubMed&dopt=Citation&list_uids=16861667

[3] Maturana H, Mpodozis J. The origin of species by means of natural drift. Rev Chil Hist Nat. 2000;261-310.

[4] Varela FJ, Thompson E, Rosch E. The Embodied Mind: Cognitive Science and Human Experience [Internet]. MIT Press; 1993. Available from: http://doi.wiley.com/10.1111/j.1468-0149.1965.tb01386.x

[5] Curtis H, Barnes N. Biologia. 2000.

[6] Merleau-Ponty M. Fenomenología de la Percepción. 1era ed. Editorial Planeta-Agostini. Barcelona; 1993.

[7] von Foerster H. Las semillas de la cibernética. 1996.

[8] Varela F. El Fenomeno de la Vida. Santiago de Chile: Dolmen Ediciones S.A.; 2000.

[9] Maturana H. La objetividad: un argumento para obligar [Internet]. Santiago de Chile: Paidos; 1997. Available from: http://www.radiomanque.org/wp-content/uploads/2017/07/Maturana-Humberto-La-Objetividad-Un-Argumento-Para-Obligar.pdf

[10] Maturana H. Desde la Biología a la Psicología. Viña del Mar, Chile: Ediciones Fundación SYNTHESIS; 1993.

[11] De Waal F. El bonobo y los diez mandamientos. Tusquets Editores S.A.; 2013. 283 p.

[12] Maturana H, Mpodozis J. Percepción: configuración conductual del objeto. 1987. p. 60-8.

[13] Maturana HR. The Organization of the Living: A Theory of the Living Organization. Int J Man Mach Stud. 1975;7(3):313-32.

[14] Gallagher S. Bodily self-awareness and object perception. 2003;VII(1995).

[15] Gallagher S. Dimensions of Embodiment: Body Image and Body Schema in Medical Contexts. In: Toombs SK, editor. Handbook of Phenomenology and Medicine. Kluwer Academic Publishers; 2001. p. 147-75.

[16] Thompson E. Mind in life: biology, phenomenology, and the sciences of mind. United States of America: Harvard University Press; 2010.

[17] Gallagher S. Body Image and Body Schema: A Conceptual Clarification. J Mind Behav. 1986;7(4):541-54.

[18] Bauer CCC, Díaz JL, Concha L, Barrios FA. Sustained attention to spontaneous thumb sensations activates brain somatosensory and other proprioceptive areas. Brain Cogn [Internet]. 2014;87(1):86-96. Available from: http://dx.doi.org/10.1016/j.bandc.2014.03.009

[19] Longo MR, Haggard P. An implicit body representation underlying human position sense. Proc Natl Acad Sci U S A. 2010;107(26):11727-32.

[20] Ehrsson HH, Kito T, Sadato N, Passingham RE, Naito E. Neural substrate of body size: illusory feeling of shrinking of the waist. PLoS Biol [Internet]. 2005 Dec [cited 2013

Jun 16];3(12):e412. Available from: http://www.pubmedcentral.nih.gov/articlerender.fcgi?artid=1287503&tool=pmcentrez&rendertype=abstract

[21] Maturana H. Comment by Humberto R. Maturana: The Mind is not in the head. J Soc Biol Syst. 1985;8(4):307-9.

[22] Lackner JR. Some proprioceptive influences on the perceptual representation of body shape and orientation. Brain. 1988;111:281-97.

[23] Hospod V, Aimonetti JM, Roll JP, Ribot-Ciscar E. Changes in human muscle spindle sensitivity during a proprioceptive attention task. J Neurosci. 2007;27(19):5172-8.

[24] Lefaivre SC, Almeida QJ. Can sensory attention focused exercise facilitate the utilization of proprioception for improved balance control in PD? Gait Posture [Internet]. 2015;41(2):630-3. Available from: http://dx.doi.org/10.1016/j.gaitpost.2015.01.013

[25] Plutchik R. The Emotions. Rowman & Littlefield inc; 1991.

[26] Damasio A. The Strange Order of Things: Life, Feeling, and the Making of Cultures [Internet]. New York: Pantheon Books; 2018.

[27] Damasio AR, Grabowski TJ, Bechara A, Damasio H, Ponto LLB, Parvizi J, et al. Subcortical and cortical brain activity during the feeling of self-generated emotions. Nature. 2000;09:1049-56.

[28] Peil KT. Emotion: The Self-regulatory Sense. Glob Adv Heal Med. 2014;3(2):80-108.

[29] Pollick FE, Paterson HM, Bruderlin A, Sanford AJ. Perceiving affect from arm movement. Cognition. 2001;82:B51-61.

[30] Shafir T, Tsachor RP, Welch KB, Allard ES. Emotion Regulation through Movement : Unique Sets of Movement Characteristics are Associated with and Enhance Basic Emotions. Front Psychol. 2016;6(January):1-15.

[31] Vásquez-Rosati A. Body Awareness to Recognize Feelings The Exploration of a Musical Emotional Experience. Constr Found. 2017;12(2):219.

[32] Koch SC, Fuchs T, Summa M. Body memory and kinesthetic body feedback: The impact of light versus strong movement qualities on affect and cognition. Mem Stud. 2014;7(3):272-84.

[33] Schmalzl L, Crane-Godreau MA, Payne P. Movement-based embodied contemplative practices: Definitions and paradigms. Front Hum Neurosci. 2014;8(1 APR):1-6.

[34] Vasquez-Rosati A, Brunetti EP, Cordero C, Maldonado PE. Pupillary Response to Negative Emotional Stimuli Is Differentially Affected in Meditation Practitioners. Front Hum Neurosci [Internet]. 2017;11(May):1-9. Available from http://journal.frontiersin.org/article/10.3389/fnhum.2017.00209/full

Chapter 8

Nomophobia Kids and Proprioception

Giridharan Vaishnavi

Abstract

Proprioception is the sense of self movement and body position. The CNS integrates proprioception and other sensory system such as vision and vestibular system in order to create body position, movement and acceleration. It is developed by movement and works with surroundings. Children using smartphones for a long time result greater impact on the sensorimotor function and their proprioception is affected. In this topic, the write up is going to be regarding the proprioceptional deficit and the problems associated because of that of children using mobile phones for a prolonged period of time. The proprioceptive system has an extensive influence at the protection of human fitness. When the proprioception is dysfunctional, the vital anxious device does no longer recognize the ideal fame of tonicity of the muscular tissues at rest or in motion, does no longer combine effectively the records that comes from sensory receptors, and has issue in modulating multi-sensory integration, with outcomes in motor behavior and cognitive function.

Keywords: nomophobia, proprioception, smart phone, feedback loop, feedforward loop

1. Introduction

In the last few years, the usage of smartphone has been progressively increased worldwide among kids. Nomophobia is described as an experience of anxiety due to fear of not having access to mobile phones. Smartphone addiction leads to restless night, anxiety, social isolation, nervous breakdown, weight changes, insomnia & anger [1]. Nomophobia is described as" the discomfort or tension because of the non-availability of a cell cellphone, non-public computer (laptop) or any other digital device" [2].

Nomophobia is described as "the soreness or tension as a result of the non-availability of a cellular telephone, personal laptop (pc) or any some other digital verbal exchange device" [2]. Clayton et al. references Belk (2013) of their explanation of smart-phone loss as the "unintentional lack of a ownership need to be regarded as a loss or lessening of self" [3].

Long time usage of smartphones by the kids lead to musculoskeletal problems due to the faulty posture maintained like forward neck posture problems, rounded shoulder or slouched posture for a prolonged period of time [4]. The structural problems caused by faulty posture may lead to decrease in proprioception there by resulting in decreased balance ability. The maintenance of proprioception is extremely important in order to prevent injuries & this is mainly subjected to proprioceptive input from mechanoreceptors in the capsule, ligament & tendon added to vestibular & visual input to central nervous system.

The proprioceptive system is part of the central nervous system. Proprioception is one's personal sensation of the body. Proprioception is the feel arising from joints, muscle groups, tendons and related deep tissues that offer statistics to the central nervous system (CNS) about static and dynamic motion of limbs and the body. Proprioception information is processed at exclusive levels of the central nervous system in order to meet physical needs positioned at the body and aides a mover to arrive at motor choices. Proprioception strongly contributes to the fitness of joint balance; sensing passive or energetic joint articulation; joint load absorption and rebound; and muscle duration, force and velocity. Abnormality and tissue trauma excluded, the proprioception is a basic part of all people's neuro-anatomy [5].

The proprioception experience end result of a collection of fact derived from sensory receptors found within, muscular tissues, tendons, ligaments and fascia. The sensory receptors answerable for proprioception are referred to as mechanoreceptors and are the subgroup of somatosensory system. All mechanoreceptor facts is grouped collectively in the central nervous system (CNS) and are processed at distinct ranges of the consistent with unique movement and environmental demands.

The proprioception information is blended with body senses on a moment by using secondary basis for the duration of real time events. The simultaneous interaction of many sensory procedures create a collective internet notion of the body in area and contributes to motion choices.

Nonpublic, bodily and perceptual sources are mixed with genetic, cultural and societal sources as contributing factors to motion or action selections. Throughout the body & nervous system proprioception plays the role of a loop which continuously feedback & forward input. The vital importance of optimal orientation and postural control is relied on the complex reflex & central interaction between cervical proprioception, vestibular & visual information [6]. When there is an increased use of smartphone by the kids the proprioception lies at the boundary between neurophysiology & neuropsychology. The nomophobia children will soon develop poor balance, poor co-ordination and increased postural sway.

From an engineering perspective the human neuro musculoskeletal machine can be appeared as a robot, consisting of a linkage system (the skeleton) with motors (muscles), sensors (proprioceptors), and a control gadget (the CNS) [7]. The nerves and neurons are the wires and connectors, which shipping the data from the proprioceptors to the CNS and from the CNS to the muscular tissues.

The CNS integrates intentions with facts from the proprioceptor to coordinate motion ofthe skeleton through selectively (de-)activating muscle groups. Manage engineers will directly recognize a feedback loop: the movement outcomes from commands from the CNS, which on their flip (in part) rely upon the movement sensed via the proprioceptors there is a mutual interaction among CNS and limb movement. Postural manipulate is a specific case inside human movement control [8].

The human has to maintain a posture, i.e. an equilibrium position. At some point of postural manipulate unbiased strategies contribute to stability and performance: (1) intrinsic residences of the muscles and (2) proprioceptive reflexes. Balance is controlled and maintained by a complex set of sensory-motor control system which include the sensory input, receptors, input from vestibular system, integration of sensory input and the motor output [9].

All children receive information from their internal and outside environments through the following senses: imaginative, motion (vestibular), prescient (ocular), hearing (auditory), taste (gustatory), odor (olfactory), contact (tactile), Joint and muscle cognizance (Proprioceptive).

The term sensory integration refers back to the manner of receiving and responding to the incoming facts. It starts off evolved while your toddler gets information from their senses, then the vital nervous device directs the records to the correct parts of the mind, and the records is integrated so the child can reply in the perfect manner.

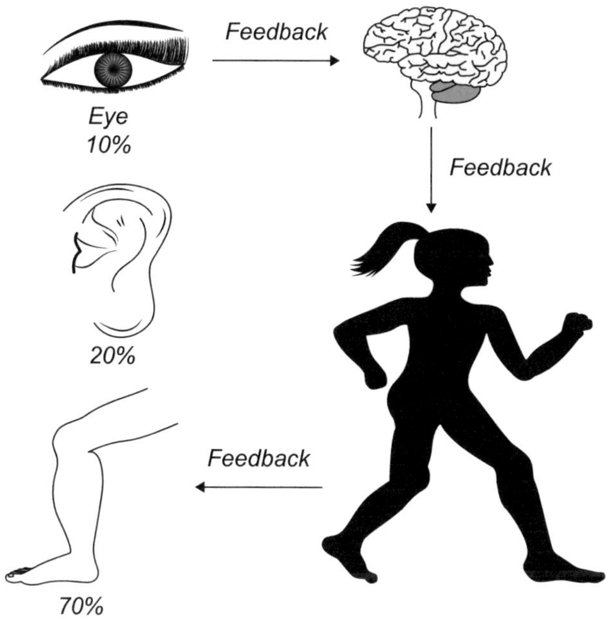

2. Proprioceptive receptors

The term proprioceptor comes from the Latin 'proprius', one's own, and 'recipio', to get. It was characterized by Sherrington (1906): 'In muscular receptivity we see the body itself going about as an upgrade to its own receptors – the proprioceptors'. Types of proprioceptors include muscle spindle, golgi tendon organs and the mechanoreceptors in the joint capsule.

When muscles lengthen, the spindles gets stretched. This stretch activates the muscle spindle that successively sends an impulse to the neural structure. This impulse leads to the activation of a lot of motor neurons at spinal level that send an impulse back to the muscle [10]. The Golgi connective tissue Organ could be a propioception receptor that's situated at intervals the tendons found on every finish of a muscle. It responds to inflated muscle tension or contraction as exerted on the connective tissue, by inhibiting any muscular contraction. Mechanoreceptors meant as primary neurons or nerve endings that reply to mechanical stimuli by firing action potentials. When mechanoreceptor receives a input, it begins to fire action potentials at raised frequency [11].

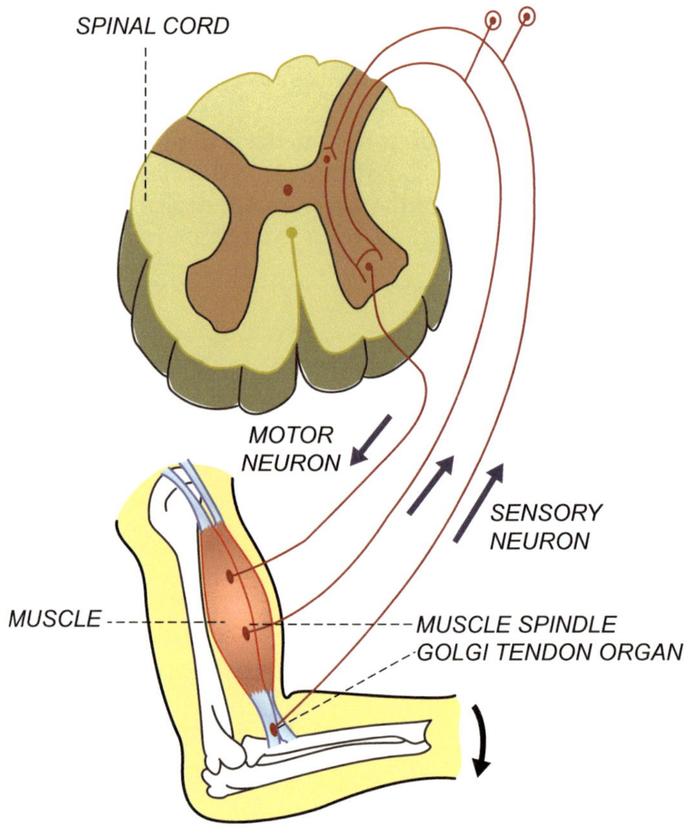

3. Proprioceptive system

Proprioceptive system has three overlapping major functions:

1. Regulation of tone

Posture and body movement is mediate by proprioception through the afferent information it receives from all sensory receptors, and by the efferent data it sends to the extrapyramidal motor tracts.

2. Egocentric abstraction localization

By integration and modulating the knowledge that comes from sensory receptors, the proprioceptive system informs the brain about the relative position of the sensory organs, the relation between every body segment, and also the relative position of the body within the close setting.

3. Modulation of multisensory data

Proprioceptive data well-known to be transmitted within the multisensorial deep layers of the nerve center within the midbrain is believed to possess a task in modulating multisensory integration. This modulation has consequences in motor behavior and better psychological feature functions [12].

DOI: http://dx.doi.org/10.5772/intechopen.96236

4. Eyes and Proprioception in nomophobia

The receptors of proprioception as major role in movement of eye control and construction of extra personal space. It was Thomas Reid in 1785 who clearly explained the mechanism of the function of eye proprioception. The visual gaze direction as direct concern with the neck muscles control during vestibular stimulation. The extra ocular muscle afferent signals are determined in patterned inhabitation of forelimb and neck muscles & there by influence the head position on the body [13].

Its miles possible to steer body proprioception by using stimulating no longer simplest direct mechanoreceptors like neuromuscular spindles, or joint or tendon neurologic terminations, however additionally through modulating the information from different sensory input as well. These range from visible receptors linked to the retinocolicular pathway in the uppermost location of the body to the only plantar receivers underfoot.

Brain will tend to combine various available source of extra-retinal signals to foster visual clarity during eye and head movements with proprioception as major contribution visual fatigue caused by usage of smartphones for a prolonged period of time as highly impact with visual ability to control posture thereby reducing the ability to balance. Thus it is very clear that continuous usage of smartphones by the kid for the sake of games & YouTube may lead to posture and balance disturbances through visual gaze [14].

5. Muscle, joints and proprioception in nomophobia

Proprioception is a chain of feedback between the sensory receptors which are located in the skin, joints and muscles when the duration of smartphones usage increases it will surely have an negative impact of cervical proprioception & dynamic balance ability.

Prolonged flexed neck, posture in turn will increases the muscle activity which will cause the musculoskeletal pain in neck and shoulder on comparison fast muscles, muscle fibers cause more fatigue than slow muscle fiber when static posture is maintained for a prolonged duration as the cervical flexion angle is decreased. When the head is inforward headed posture the muscular tissue perform cervical extension in the back of the neck are contracted isometrically creating a force which is against gravity, which will lead to prevention of cervical flexion or forward head movement, & long term isometric contraction of the muscles in the back of neck involved in the extension of cervical & can also cause pain by stimulating trigger points. The combination of extension within the higher cervical region and flexion within the lower cervical region seems in patients with forward head posture attribute to head posture. Changes in the cervical region by sustained poor head posture, cause excessive joint and muscle loading, and later on influencing weakness of the deep cervical muscles. Among several body structures set within the cervical region, the muscle is thought to be a main part for position sense through its receptors, like muscle spindles. The cervical vertebrae contribute proprioception sense input [15].

The proprioceptive sensing of the cervical vertebrae transmits data to correct arrangement and plays a crucial role in bodily property management. Additionally, it reacts sensitively to the fine movement of the top by acting in coordination with sensing from the vestibular apparatus. Asymmetrical alignment of the top and neck ends up in errors within the data received as visual and proprioception sensing this eventually reduces the balance and will increase the chance of falling and contractile organ injuries whereas acting activities.

6. Vestibular system and proprioception in nomophobia kids

The central and peripheral vestibular organs, in conjunction with the visual and somatosensory systems, are responsible for balance, equilibrium, and orientation in space. Vestibular system as major role in subjective awareness of body position and movements in space, postural tone and equilibrium and stabilization of the eyes in space during head movement along with visual system and proprioception continuous use of smart phones may lead the child hyper expressive and also lead to gravitational insecurity, which is an sensory integration issue which may cause to react to movements in an extreme way. This is due to the impact of vestibular system by which the child's gravity receptors become extra sensitive.

7. Motor control and proprioception in nomophobia kids

Proprioception is caused out to all the levels of central nervous system where it gives out unique sensory component to effectively use motor control neuromuscular control of dynamic restraints very necessarily needs proprioceptive information. The motor control is disturbed in nomophobia kids through stress imposed on other bodily system. Perception is extraordinarily necessary in control as a result of it carries the relevant info concerning objects, environments and bodies that is employed in organizing and initiating actions and movements.

proprioceptive data plays an integral role within the ability to change internal models used with feed forward management that has been incontestible to be solely part paid for by visual information. The planning of movements additionally needs attention to environmental constraints. Children using smartphone for a prolonged period of time lack attention thereby their planning and control of movements is not proper.

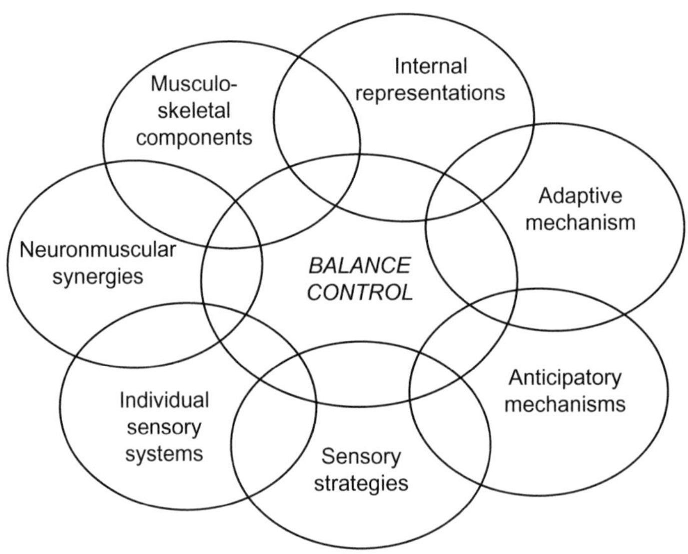

8. Proprioception impairment in nomophobia kids

- Balance Issues
- uncoordinated Movement
- Clumsiness
- Poor Postural Control
- Trouble Recognizing Your Own Strength

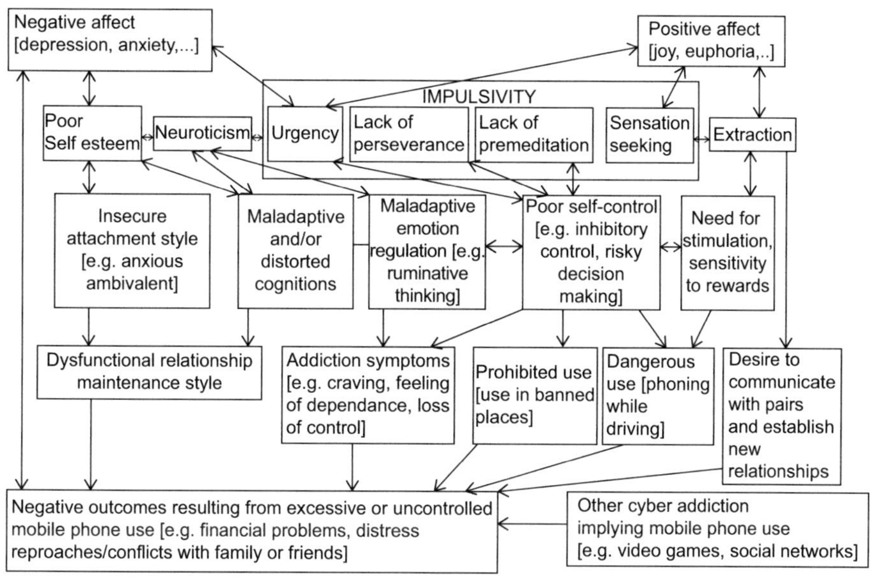

- Avoiding Certain Movements Or Activities

9. Proprioception assessment

There are a few test Physiotherapists can use to assess proprioception, depending at the body element being assessed. That includes:

Heel-shin:

The patient is asked to the touch the heel of 1 foot to the opposite knee after which to drag their heel in a instantly line all of the way down the front in their shin and lower back up once more. With a view to get rid of the impact of gravity in moving the heel down the shin, this test have to continually be done in the supine position [16].

Finger-nose-finger:

The subject is asked to alternately touch their nose and the examiner's finger as fast as possible [3].

Distal proprioception test:

The tester will be made to circulate the joints of the hip, knee ankle and big toe up and down as it is watched. Then ask the subject to repeat the equal movement together with eyes closed.

A contralateral joint matching task:

Asking the affected person to match a verified joint angle, and measuring the distinction among the real joint angle, and the reproduced joint perspective.

Romberg's test:

The affected person is requested to remove his/her footwear and stand with two feet together. The clinician asks the patient to first stand quietly with eyes open, and ultimately with eyes closed. The Romberg test is scored by counting the seconds the person is able to stand with eyes closed [3].

Spinal Motion Apparatus:

This procedure developed by Pankhurst and writer for assessing the proprioception of lower back. It is composed of a motor operated device that produces passive motion of lumberspine in three planes whereas the trunk stayed fastened. The apparatus detects motion and identifies the neutral position and the direction of movement. It assesses movement in 3 planes as advantage, but the employment during a clinical population might not be possible because it utilizes the advanced equipment [17].

Active Movement Extent Discrimination Device:

Developed by Hobbs to assess lumbar proprioception. It depends on discriminating the position differences in 11–19° of lumbar flexion. It consists of free standing with stopper at five preset distances. The subject had to discriminate preset trained flexion positions while standing. The test's disadvantage is that the subject's head is also moving through the test so the vestibular system might be adding to the proprioception sense [18, 19].

Cervicocephalic Kinesthesia:

Kristjansson et al. described the test. It has fast track sensors. Various uses of the test described such as relocation of the head to the natural position after active turn to left and right or active relocation to 30° turn from the natural head position. Passive trunk rotation of 30° or figure of eight motion can also be used before subjects repositioning head to a natural position [8].

Limb Position Copying and Reproducing Tests:

Described by Kaplan [7]. This test can be used for assessing the proprioception of various joints such as knee or elbow. The test requires active reproduction of ipsi- and contralateral positions of the limb. Goniometer measures the error between reproduction and the target.

Precaution to overcome nomophobia in kids:

- Delete all the social media apps on mobile phones.
- Set specific boundaries for usage of smartphones.
- Lock the smartphones with long password.
- Parents spend time with kids playing with them.
- Parents avoid smartphones before kids.

10. Intervention

Chair based exercises

- Hand pushes
- Hand pulls
- Head compressions
- Chair push ups
- Theraband on chair
- Squeezing a stress ball

Classroom based exercises

- Wall pushes
- Push ups
- Lifting weights e.g. tins, books, dumbells
- Jogging on the spot
- Star jumps
- Bouncing on therapy/exercise ball

Other exercises

- Climbing wall bars/ropes
- Throwing/catching weighted ball
- Crawling obstacle course
- Wheelbarrow walks
- Gymnastics- handstands, cartwheels, using gym equipment
- Jumping e.g. hop scotch
- Tug of war
- Bouncing on space hopper
- Lying on stomach over exercise ball and weight-bearing through arms

Functional activities

- Wiping benches and tables
- Brushing/mopping floors
- Holding doors open
- Carrying piles of books
- Carrying a backpack with a heavy item in it
- Stacking chairs
- Moving furniture

Proprioception ativity using swiss ball

- Roll it up on a wall.
- Dribble it. Pushing the ball into the ground is great for the proprioceptive sense as well as when it bounces back.
- Bounce on it alone.
- Kick it against a wall.

INITIAL PHASE – The first 3 weeks	FINAL PHASE – The last 3 weeks
EXERCISE 1 Initial phase: From the position indicated in the picture, to do shoulder flexion – extension. 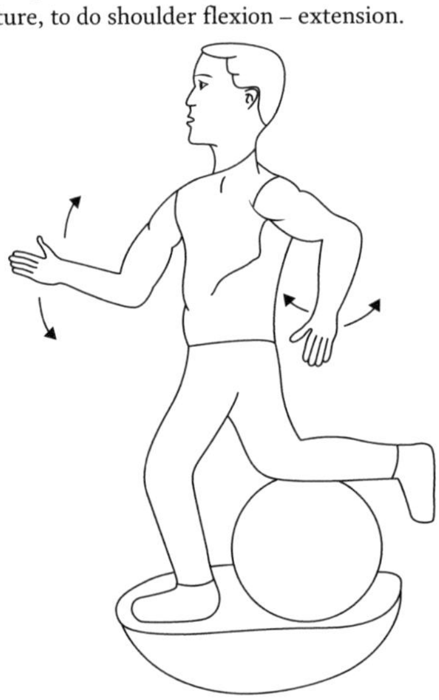 30″ each limb	**EXERCISE 1** Final phase: the same performance but now each hand holds a 2 kg weight which increases 1.5 kg per week.  30″ each limb

INITIAL PHASE – The first 3 weeks	FINAL PHASE – The last 3 weeks
EXERCISE 2 Initial phase: From the position of the picture, doing hip flexion – extension at the same time that moving the shoulder in flexion – extension. 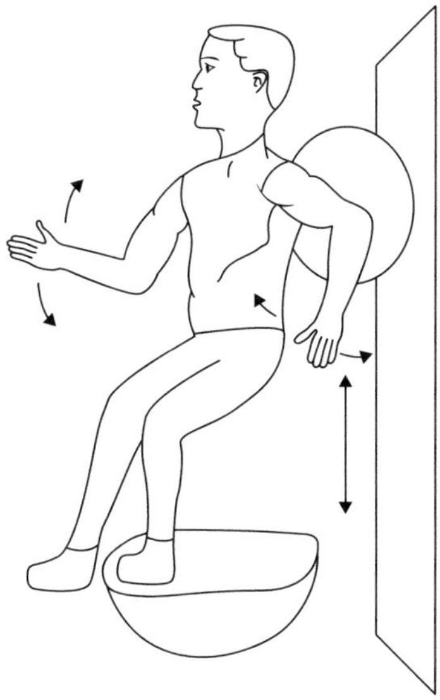 10 times each limb	**EXERCISE 2** Final phase: The same performance apart from each hand hold a 2 kg weight which increases 1.5 kg per week and a 3 kg ankle weight in each ankle. 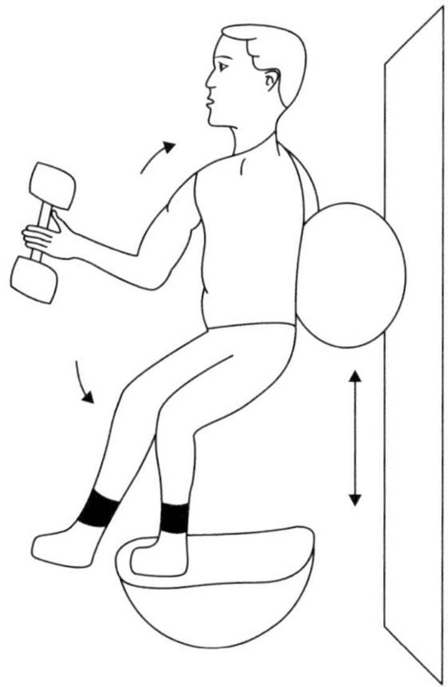 10 times each limb
EXERCISE 3 Initial phase: From the position indicated in the picture, to do shoulder flexion – extension. 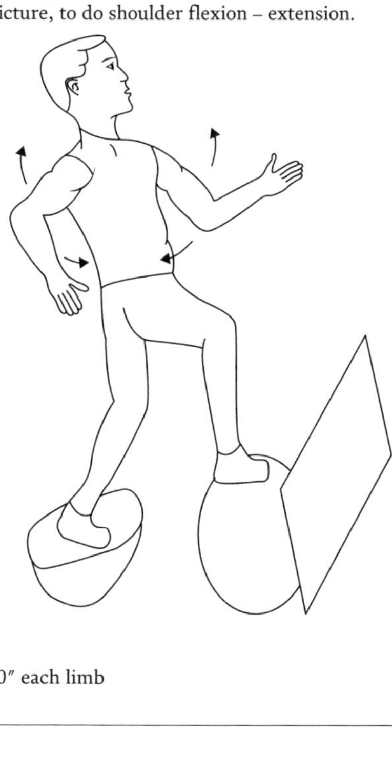 30″ each limb	**EXERCISE 3** Final phase: The same performance but now each hand hold a 2 kg weight which increases 1.5 kg per week and the back leg is supported over the metatarsals.  30″ each limb

INITIAL PHASE – The first 3 weeks	FINAL PHASE – The last 3 weeks
EXERCISE 4 Initial phase: The free leg does a whole circulation which ends with the extension hip. 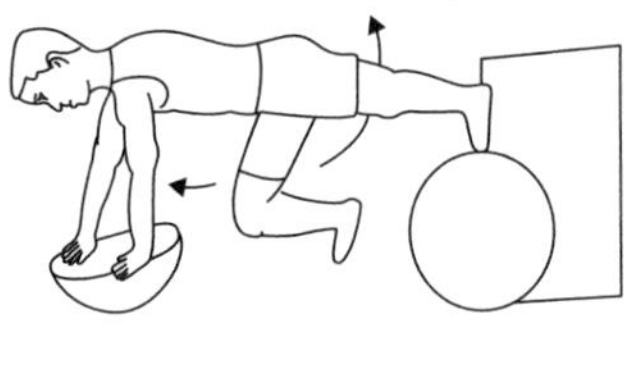 10 times each limb	**EXERCISE 4** Final phase: The same performance apart from a 3 kg ankle weight in the free leg. 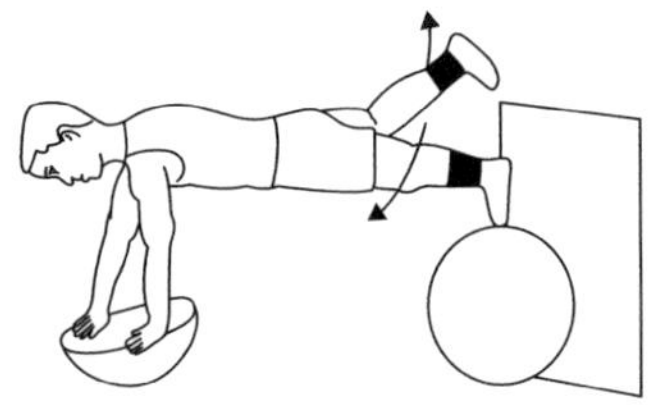 10 times each limb
EXERCISE 5 Initial phase: The free leg does hip, knee and ankle flexion synchronized with the high member which also moves in flexion – extension. 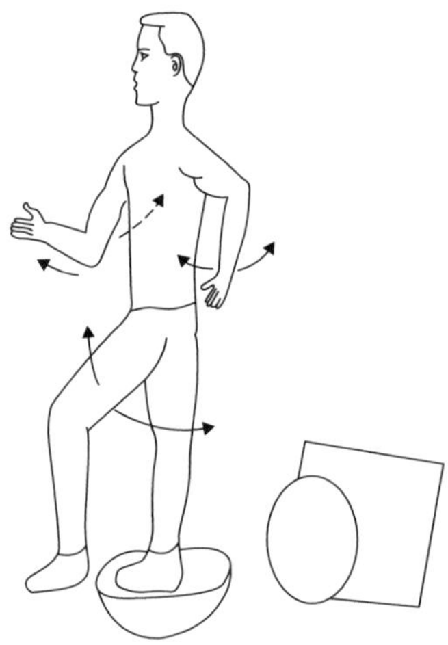 10 times each limb	**EXERCISE 5** Final phase: The same performance apart from each hand hold a 2 kg weight which increases 1.5 kg per week and also a 3 kg ankle weight in each ankle. 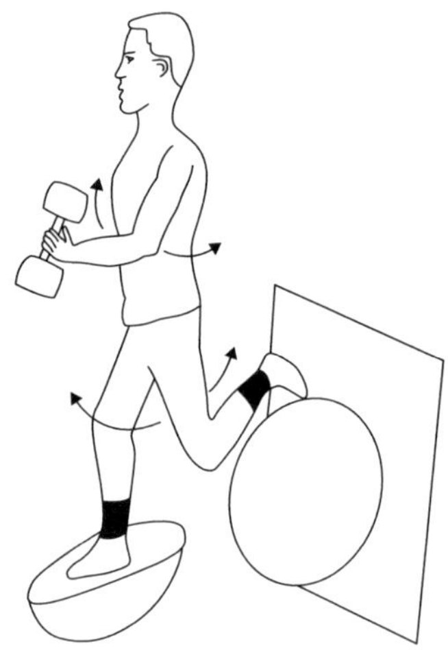 10 times each limb

11. Conclusions

In line with King et al. [2], this study purports that nomophobia, or no mobile phone phobia, is thought of a contemporary age phobia introduced to our lives with the speedy proliferation and adoption of smartphone.

The proprioceptive system has an extensive influence at the protection of human fitness. When the proprioception is dysfunctional, the vital anxious device does no longer recognize the ideal fame of tonicity ofthe muscular tissues at rest or in motion, does no longer combine effectively the records that comes from sensory receptors, and has issue in modulating multi-sensory integration, with outcomes in motor behavior and cognitive function. This outcomes in a wide variety of proprioceptive abnormalities which are clinically related and are handled collectively termed as Postural Deficiency Syndrome (PDS). Kids using smart phones for a prolonged period of time lack attention, lack posture and motor control. Smart phone usage for extended period may change the brain activity, and postural disturbance.

Acknowledgements

I would like to thank the authorities of Dr. MGR Educational and Research Institute, University, Honorable chairman Dr. AC. SHANMUGAM, Honorable president ACS. ARUNKUMAR, Honorable secretary Thiru A. RAVIKUMAR and our Principal, Faculty of Physiotherapy Dr. CV. SENTHIL NATHAN for providing facilities required to complete the chapter with their support and guidance. With great privilege I thank my husband Er. S Elangovan and my mother for their support and motivation.

Author details

Giridharan Vaishnavi
Faculty of Physiotherapy, Dr. MGR Educational and Research Institute, Chennai, India

*Address all correspondence to: vaishnavigiri2012@gmail.com

References

[1] Geser H. Toward a sociological theory of the mobile phone.Zürich: Soziologisches Institut der Univ. Zürich: Online Publications 2004 [cited 2011 Sept 26]. Available from:http://socio.ch/mobile/t_geser1.htm

[2] King, Valença, Silva, Sancassiani, Machado, & Nardi, Nomophobia: impact of cell phone use interfering with symptoms and emotions of individuals with panic disorder compared with a control groupClin Pract Epidemiol Ment Health 2014 Feb 21;10:28-35. doi: 10.2174/1745017901410010028. eCollection 2014.

[3] Gokben Hızlı Sayar, Huseyin Unubol, Assessing Proprioception, The Journal Of Neurobehavioral Sciences, 10.5455/Jnbs.1485955027, Volume 4 / Number 1 / 2017

[4] Miranda Janeschild, BS, MS, OTR/L, CFP, The Role Of The Proprioceptive System In Learning Dance While Studying The Axis Syllabus of sensitivity of proprioception by a new clinical test: The dual joint position test. Clinical Neurology and Neurosurgery, 115(7), 1023–1027. http://doi.org/10.1016/j.clineuro.2012.10.017

[5] Bevan, L., Cordo, P., Carlton, L., & Carlton, M. (1994). Proprioceptive coordination of movement sequences: discrimination of joint angle versus angular distance. Journal of Neurophysiology, 71(5), 1862–1872.

[6] Bhanpuri, N. H., Okamura, A. M., & Bastian, A. J. (2012). Active force perception depends on cerebellar function. Journal of Neurophysiology, 107(6), 1612–1620. http://doi.org/10.1152/ jn.00983.2011

[7] Kaplan, F. S., Nixon, J. E., Reitz, M., Rindfleish, L., & Tucker, J. (1985). Age-related changes in proprioception and sensation of joint position. Acta Orthopaedica Scandinavica, 56(1), 72–74. http://doi.org/10.3109/17453678508992984

[8] Kristjansson, E., Dall'Alba, P., & Jull, G. (2001). Cervicocephalic kinaesthesia: reliability of a new test approach. Physiotherapy Research International: The Journal for Researchers and Clinicians in Physical Therapy, 6(4), 224–35. http://doi.org/10.1002/pri.230

[9] Alves da Silva T (2017) Low-powered asymmetric prismsimprove eye-hand coordination in patients with PosturalDeficiency Syndrome.

[10] Roll JP RR (1988) From eye to foot: a proprioceptive chaininvolved in postural control. Elsevier, Amsterdam

[11] Roll R, Velay JL, Roll JP (1991) Eye and neck proprioceptivemessages contribute to the spatial coding of retinal inputin visually oriented activities. Exp brain Res 85:423–431.

[12] Adams, J. A. Feedback theory of how joint receptors regulatethe timing and positioning of a limb. *Psychological Review,* 1977, 84, 504-523.

[13] Adams, J. A. Theoretical issues for knowledge of results. In G. E. Stelmach (Ed.), *Information processing in motor control and learning.* New York: Academic Press, 1978.

[14] Deepika Singla and Zubia Veqar. Association Between Forward Head, Rounded Shoulders, and Increased Thoracic Kyphosis: A Review of the Literature. J Chiropr Med. 2017 Sep; 16 (3): 220–229.

[15] Asatryan, D. C., Fel'Dman, A. D. Functional tuning of thenervous system with control of movement or maintenance of asteady posture. I. Mechanographic analysis of the work on

thejoint on execution of a postural task. *Biophysics*, 1965, 10, 925-935.

[16] William S. Marras - Marras William S, Davis Shelby W, Miller Robert J, Mirka Gary A Apparatus For Monitoring The Motion Components Of The Spine*Grant Us-5143088-A - Published 1992-09-01*

[17] Susan Antcliff, Marijke Welvaert, Jeremy Witchalls, Sarah B. Wallwork, Gordon Waddington. Using the Active Movement Extent Discrimination Apparatus to Test Individual Proprioception Acuity: Implications for Test Design, December 3, 2020

[18] G.Vaishnavi, S. Reena Mary, S. Shanthi Santhosh Kumari, R. Deenu Priya, Archana R. Menon, Clitty, Correlation of Mobile Phone Addiction Scale (Mpas) with the Craniovertebral Angle and Beck's Depression Inventory in Bike Drivers and Students, NOVYI MIR Research Journal, Volume 5, Issue 6, 2020-2173

[19] Russell B. Clayton, Glenn Leshner, Anthony Almond, The Extended iSelf: The Impact of iPhone Separation on Cognition, Emotion, and Physiology, Journal of computer – mediated communication 08 January 2015 https://doi.org/10.1111/jcc4.12109

Chapter 9

Proprioception in Immersive Virtual Reality

Alexander Vladimirovich Zakharov, Alexander Vladimirovich Kolsanov, Elena Viktorovna Khivintseva, Vasiliy Fedorovich Pyatin and Alexander Vladimirovich Yashkov

Abstract

Currently, in connection with the advent of virtual reality (VR) technologies, methods that recreate sensory sensations are rapidly developing. Under the conditions of VR, which is an immersive environment, a variety of multimodal sensory experiences can be obtained. It is urgent to create explicit immersive environments that allow maximizing the full potential of VR technology. Activation of the proprioceptive sensory system, coupled with the activation of the visual analyzer system, allows you to achieve sensations of interaction with VR objects, identical to the sensations of the real physical world. Today, the activation of proprioceptive sensations is achieved using various devices, including robotic ones, which are not available for use in routine medical practice. The immersive multisensory environment makes it possible to significantly personalize the rehabilitation process, ensuring its continuity and effectiveness at various stages of the pathological process and varying degrees of severity of physical disorders, while significantly reducing the burden on the healthcare system by automating the rehabilitation process and objectively assessing the effectiveness. Further development and increased availability of VR technologies and devices that allow achieving an increase in immersion due to sensory immersion will be in great demand as a technology that allows teaching patients motor skills.

Keywords: virtual reality, proprioception, neuroplasticity, physical rehabilitation, habilitation

1. Introduction

Proprioception is the sensation of the position of body parts relative to each other in space, both in statics and during their movement. The formation of proprioceptive sensation occurs due to the activity of various receptor systems located in the tissues of the human body, the largest number of which is in the muscular system. Proprioception belongs to the somatosensory system and is traditionally defined as sensory sensations of position, movement, or balance [1]. From this position, proprioception is the awareness of the position or movement of our body and its parts in space because of processing information from the receptors of muscles, joints, tendons, and skin. This type of sensory sensation includes two

IntechOpen

components, the first of which arises in a static position, and the second appears during movement and plays a decisive role in ensuring coordination of movements; both are important in ensuring balance at rest, as well as during movement [2].

The proprioceptive system has a high degree of precision in registering changes recorded by the receptor apparatus, more specifically, a change in the angle in a particular joint or in the mass of a held object. In the process of ontogenesis, the role of proprioception as the dominant receptor system in sensory cognition of the external world goes into second place relative to the visual analyzer, which in humans and mammals acquires a dominant role. However, proprioception continues to be in high demand in building a complete judgment about an external object, providing information about its physical properties necessary to ensure effective interaction with this object, which underlie more complex motor skills that are important for the formation of praxis. Proprioception is actively used as an afferent system for the formation of biofeedback when adjusting the motor system to external conditions in the process of performing a movement and object interaction at the level of the spatial field.

The use of devices for activating proprioceptive sensations as a way to increase sensory immersion in VR or, in another way, to achieve a greater degree of immersion, is one of the unresolved, but actively developing ways to expand the practical application of VR. In VR conditions, sensory perception of reality only through the visual analyzer does not provide the formation of a complete sensory sensation, identical to physical interaction with objects, and cannot be used to implement explicit interaction. It is precisely because of the difficulty of achieving such a quality of proprioceptive sensations, which would be identical to the sensations received in the physical world, that modern VR systems are mostly implicit, and control in them is implemented in surrogate ways that are not identical to the natural richness of object sensations and the complexity of manipulating them.

At the moment, less attention is paid to the study of unimodal proprioceptive information processing in statics or only when performing a movement than the study of multisensory integration processes, with the participation of not only the proprioceptive system, but also other sensory analyzers. Isolated activation of the proprioceptive system is possible only when the visual analyzer is not functioning, which is rarely observed under normal conditions. The close connection between the proprioceptive system and the visual analyzer can be traced through experiments on the formation of proprioceptive sensations, without directly affecting the receptor apparatus. Manipulation of proprioceptive sensations is possible using a visual analyzer, as demonstrated in experiments on the use of mirror therapy in the treatment of phantom pain after limb amputation due to various reasons. In this case, through the information coming through the visual analyzer, it is possible to achieve proprioceptive sensations, sensations of movement, as well as a sense of touch in the complete absence of somatosensory stimuli.

This is evidence that proprioception is a complexly organized bodily sensation, the formation of which can be influenced by the activation of the receptor apparatus of various sensory systems and, of course, primarily the visual analyzer.

2. Possibilities of using VR in physical rehabilitation

One of the main and important applications of VR is its use in the medical field to provide various tasks of physical rehabilitation and habilitation.

The main goal of physical rehabilitation is to help a person return to a natural state when performing daily activities by restoring damaged motor skills, as well as if they cannot be completely restored, acquiring new ones that compensate for those lost due to diseases of the musculoskeletal system (trauma, pathology of the muscular system) or the CNS vascular diseases and injuries. Also, significant prospects for the use of VR are traced in patients with impaired formation of the motor system in ontogenesis.

Research on the effectiveness of rehabilitation using VR technology appeared in the mid-1990s.

Currently, there are several systematic reviews evaluating the clinical efficacy of sensorimotor training in VR in restoring the function of the upper limbs and gait after stroke [3–6].

Despite the fact that most often the study of the effectiveness of various methods of motor rehabilitation occurs on the example of such pathologies as acute disorders of cerebral circulation and traumatic brain injury due to their widespread prevalence, the study of the effectiveness of rehabilitation in other pathological conditions using VR technology is also carried out. It should be noted that the nature of movement disorders in various pathologies is accompanied by impairments at various levels of its organization, therefore, modeling various motor tasks under VR conditions allows one to obtain positive results on the restoration of movement in patients with disorders such as infantile cerebral palsy and multiple sclerosis, which are characterized by impaired organization of the motor system at various levels from the cortical to the level of paleokinetic regulation (rubro-spinal) [7]. However, these studies are sporadic and do not allow for a systematic analysis.

Thus, VR technology has significant prospects both in the rehabilitation of patients with various dysfunctions of the CNS and musculoskeletal system in the framework of restoration of function within the framework of physical rehabilitation, and in the development of unformed motor skills in the process of solving the problems of motor habilitation.

The first reason that makes VR a promising environment for solving a complex of problems of physical habilitation and rehabilitation is that VR can be used to ensure interaction with the outside world by patients with pronounced motor or other limitations [8]. The second important factor is that the VR environment and interaction with its objects can be adapted to the patient's existing physical defect, achieving significant personalization, which, accordingly, will contribute to the achievement of a more significant effect of rehabilitation or habilitation.

According to the data obtained when assessing the effectiveness of restoration of the motor function of the upper limb in patients after acute cerebrovascular accident in comparison with the group of patients receiving only traditional methods of motor rehabilitation, the use of VR rehabilitation showed great effects both in the short and relatively long term (for example, 3 months after the onset of pathology) [9].

A significant advantage of using VR is the ability to automate the rehabilitation process, use the autotune of exercises, obtain objective analytics in the rehabilitation process, and reduce the burden on rehabilitators. If we consider VR from this side, then it can be characterized as a technology that is an interface between the user and various technical devices, which makes it possible to simulate a wide variety of rehabilitation or habilitation environments, which is necessary to solve a whole range of tasks, allowing the rehabilitated to interact with objects of the VR environment through a variety of sensory channels.

Thus, the main tasks and prospects of using VR are quite clear, but at the same time they have some limitations in achieving these advantages, relative to traditional methods of physical rehabilitation or habilitation.

New approaches to rehabilitation, habilitation or training emerging on the basis of VR are based on the latest technological advances, including the use of robotic devices, tactile interfaces, and brain-computer interfaces. So, the variety of technical devices allows you to more effectively use the capabilities of VR, as an interface between the patient and the outside world, in which you can simulate various conditions and tasks, thus achieving a personalized approach to solving rehabilitation problems [10]. In addition to restoring motor skills, VR can be considered as a tool in the treatment of cognitive impairments of various origins, in the treatment of post-traumatic stress disorder, as well as various pain syndromes. One of the promising options for the use of VR is its use in the study of the ontogeny of the nervous system, which will be extremely necessary for understanding the formation of a pathological process, since any pathology in the context of ontogeny can be considered as regressive development, i.e. return to earlier stages of the existence and functioning of the CNS as a whole or its individual components [11]. In this regard, it is extremely important to provide neuro-feedback, preferably multisensory, primarily through the visual and proprioceptive sensory channels, because the formation of motor skills without these sensory systems is not possible. In several studies on the adaptation of the motor system after the demonstrated visual inconsistency of movements in VR to movements in the real physical world, it was found that the activity of the visual-motor connection in children is higher than in adolescents and adults, since such adaptation to motor activity in real in the physical world, after such a demonstration in children, it was much slower [12, 13]. Over time, as they grow older, this connection weakens, but obviously, only because of functional restructuring, and not anatomical, therefore, in a saturated immersive environment, you can get the results of visual-motor interaction, which are not always achievable in the physical world. Thus, a separate study of the motor system, as well as its functioning in health and disease, without connection with sensory systems, is inappropriate and incorrect from a physiological point of view.

VR technology can be used to provide a meaningful and effective impact on various human sensory systems; it also provides significant opportunities for modeling their functioning to compensate for lost sensory and motor functions. Now, the most technically simple VR systems are with implicit interaction with VR objects. These VR environments allow the person to be rehabilitated to act as a passive observer of the displayed content, while interaction occurs through the visual and auditory channel. Some expansion of the application of this technology occurs due to the use of various manipulators or joysticks, while the interaction with VR objects acquires the features of explicit interaction, however, the manipulation of VR objects with the use of these VR devices by objects is surrogate in nature, not giving the fullness of physical sensations.

The main problem in this interaction is the absence of physical sensations provided by the proprioceptive system, namely, the feeling of weight, density of an object, position in space. The importance of these sensations in the process of rehabilitation or habilitation lies in the fact that most of the movements performed by a person are the result of the functioning of the proprioceptive system, which ensures all their diversity and successful performance, regardless of external and internal factors. Thus, locomotor activity, devoid of fine tuning, provided by biofeedback based on the proprioceptive system, becomes less effective, and in some situations it is simply impossible, for example, walking at night or interacting with an object in three-dimensional space in the absence of visual control in patients with afferent paresis in the hand. An exception is interaction with objects in three-dimensional space, which is provided by cortical motor centers and the corresponding centers of praxis, in the implementation of which the visual analyzer has an equal or more significant role.

3. Prospects for VR in motor rehabilitation

Understanding the importance of ensuring the formation of proprioceptive sensations, when solving a range of physical rehabilitation tasks, is also based on a more intense activation of neuroplasticity processes when performing motor tasks with multisensory and, first of all, proprioceptive reinforcement, demonstrated in a number of experiments both on biological models and in patients with different pathology. The formation of biofeedback stimulates the processes of neurogenesis and neuroplasticity, due to the formation of new interneuronal connections, initially of a functional nature, but subsequently fixed at the structural level due to the activation of latent connections between individual functional structures of the CNS. That is why VR is an ideal immersive environment that makes it possible to maximally activate the processes of neuroplasticity for more effective recovery of not only motor, but also cognitive impairments in patients with damage to the CNS of various origins.

Learning how to perform new skills within the framework of solving motor tasks modeled in VR is of decisive importance for inducing neuroplasticity processes, and as a result, contributes to a more effective restoration of impaired functions due to various injuries of the CNS or the musculoskeletal system. The immersive environment created in VR is a particularly effective tool for carrying out tasks of interacting with objects in three-dimensional space. The restoration of this level of motor function is especially important in increasing the independence of patients with motor defects of varying severity, which is one of the ultimate goals of motor rehabilitation or habilitation. Additional technical capabilities in the form of using various sensors, telemetry during training in VR allow a detailed analysis of the motor activity of the rehabilitated person, which can serve as initial data for the formation of a recommendation system to increase the effectiveness of the rehabilitation process, or to adapt exercises during the rehabilitation process, for example, by gradually complicating them. Implementation, not allowing the rehabilitated person to lose interest and motivation to practice.

For example, today there is many studies that formed the basis of several publications demonstrating the effectiveness of restoration of upper limb functions in patients after acute cerebrovascular accident when using exercises with VR [14].

However, these studies were based only on clinical data carried out in patients in the late rehabilitation period after suffering acute cerebrovascular accident. At the same time, the high safety and effectiveness of such exercises in VR suggests that the use of virtual reality will also be effective in patients in the acute period, after acute cerebrovascular accident. Such results regarding the use of implicit multimodal VR with visual and proprioceptive confirmation of walking have been demonstrated in a study on the restoration of lower limb function in patients in the acute period of stroke **Figures 1** and **2** [15].

One of the reasons underlying the effectiveness of using VR as a method for restoring motor function is the ability to model new motor tasks that make the rehabilitation process more interesting, increasing the motivation of patients for further exercises [16].

VR can be used for multimodal sensory impact on the rehabilitated person. The addition of multimodal sensory reinforcements after performing the required interactions with virtual objects made it possible to use VR in a wide variety of areas, and it also significantly increases the potential for using this immersive environment in motor rehabilitation. Solving the problems of motor rehabilitation, using personalized, motor training, has a more significant impact on the processes of neuroplasticity than implicit or extra-contextual interaction with objects of the VR environment. For example, according to fMRI data, this method demonstrates a higher degree of activation of the motor cortex when performing specific motor tasks and when solving the problem of restoring motor disorders in patients after a stroke [17].

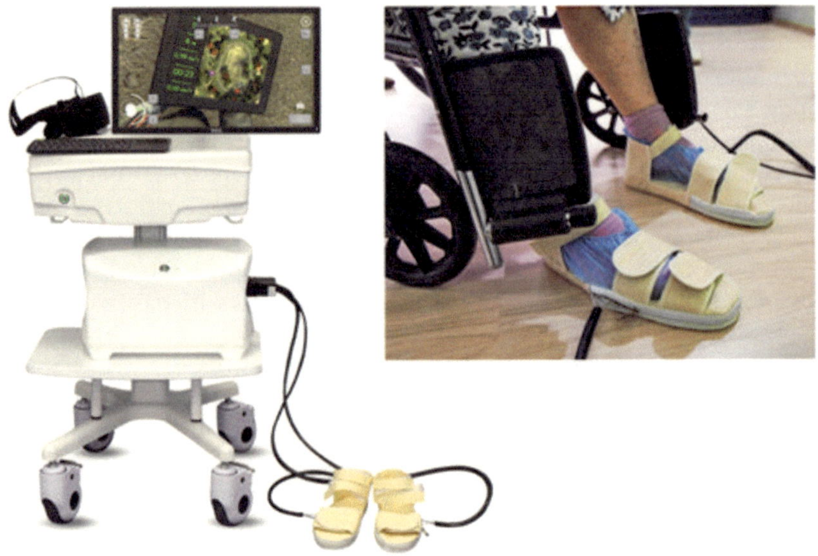

Figure 1.
ReviVR rehabilitation walk simulator.

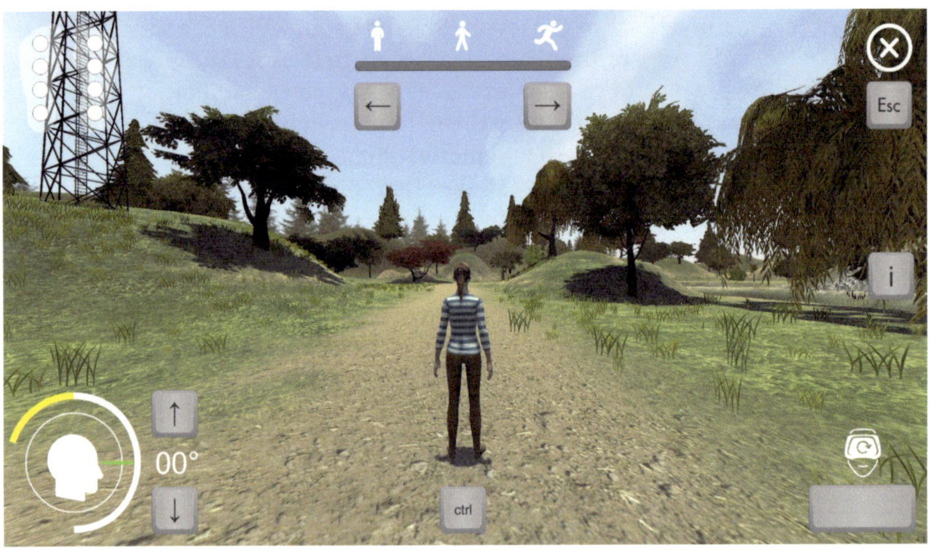

Figure 2.
First-person view in ReviVR rehabilitation walk simulator.

The data of objective methods for assessing changes in the activity of the cerebral cortex demonstrate the relationship between the specificity of the performed motor task and the degree of activation of neuroplastic changes.

Visual-motor and proprioceptive feedback, implemented in VR, provides realistic, up-to-date information during the rehabilitation exercise. It is realism and maximum proximity to physical sensations that are the most important factors that activate neuroplastic processes in the central nervous system (CNS).

Visual information, which is the most powerful sensory signal that is activated in the immersive environment, is a modeling factor for the reorganization of sensorimotor connections. For example, errors demonstrated during visual accompaniment of motor tasks performed in VR affect the motor and premotor cortex during motor learning, changing the activity of these zones [18–22].

Active and rewarding exercises (by demonstrating the progress of the performed exercise or another method) within the framework of the performed rehabilitation tasks can significantly enhance biofeedback, leading to a significant decrease in the number of errors in the restored movements, i.e. making them more energy efficient and accurate. According to fMRI data, at this moment there is a significant activation of the motor and premotor regions of the frontal cortex of the brain [23].

The very observation and subsequent ideomotor presentation of this movement leads to a significant facilitation of the formation of motor evoked potential and increases intercortical interactions in the motor and premotor regions [24–27].

It should be noted that the implementation of all this activation of the motor areas through exposure, for example, on the visual analyzer, becomes possible due to the proven numerous intrahemispheric corticocortical connections [28, 29]. These connections combine the visual cortex with the motor, premotor, parietal, and frontal lobes into a single functional system [30–34]. At the same time, there is a large number of experimental studies that demonstrate that a significant number of neurons in the motor, premotor and parietal regions can be modulated by the activity of the visual cortex of the brain [35–39].

Moreover, in contrast to proprioception, the activation of which in the physical world is necessarily associated with active or passive movements of the limbs, visual neurofeedback in VR can be provided independently of the fact of movement, for example, by simply demonstrating it. Also, it is interesting that the demonstration of this movement can be significantly changed and, most importantly, it can be completed to its full volume, regardless of the initial motor activity of the person being rehabilitated [40].

Thus, visual biofeedback allows modulation of the motor system, without the need for active or passive movements. The visual system has a high degree of reliability and specificity in the implementation of this biofeedback, because visual afferentation predominates over other sensory modalities, such as proprioceptive or auditory, and is used by a person more effectively in everyday activities [41].

An additional, but important rationale for the advisability of using the visual cortex as a sensory input for modulating motor function is that during an acute cerebrovascular accident, it is not damaged simultaneously with the motor or premotor cortex, due to their location in different blood supply basins of the brain, but namely carotid and vertebrobasilar. For acute cerebrovascular accident, in the first episode, the defeat of two pools is not a common manifestation of the disease. The defeat of the cortical representation of the visual analyzer in the form of hemianopsia and contralateral hemiplegia is observed only in the villous artery syndrome, but the preservation of the opposite visual fields allows using the VR environment for rehabilitation exercises.

Thus, VR allows the user to receive multimodal sensory information, which can cause a real sense of presence and provide cognitive, sensory and emotional immersion in the formed rehabilitation task, which has varying degrees of complexity [42–44].

The use of VR makes it possible to implement various modifications of the displayed object or its movement, highlighting it against the general background, for example, by changing the color, brightness or its shape. This opportunity allows the patient to focus on the target elements of the rehabilitation exercise, enhancing his motivation. With the help of VR, it is possible to achieve modeling of the conditions that in traditional therapy are carried out by limiting the movement of a hand that does not have motor impairments due to stroke, through its fixation to the trunk. To implement this type of therapy in VR, one can ignore the activity of a healthy limb (recorded by telemetry or contact sensors: electromyography, accelerometers, etc.) and not provide visual information regarding its movement [4, 14, 45–47]. Additional opportunities are provided by the use of the "brain-computer" interface based on the motor imagery paradigm and the P300, the use of which allows visualizing the movement of a limb with motor impairments when activity appears according to electroencephalography or functional near-infrared spectroscopy data recorded globally, with all scalp surface of the head, or only in specified areas, which are a projection onto the scalp surface of the head of the motor or premotor areas of the cerebral cortex. For example, the target signal can be used for classification within the brain-computer interface in the contralateral motor or premotor cortex, which may slow down the rate of onset of the "rehabilitation plateau" and increase the rehabilitation potential in patients with CNS pathology.

In the detected functional improvements obtained as a result of motor rehabilitation, sensorimotor activation was observed not only in the contralateral hemisphere, but also in the ipsilateral hemisphere, which indicated the activation of latent connections that were not active before the start of the rehabilitation measures [48–50]. The ongoing rehabilitation in VR and the progress obtained with it in the restoration of motor function are primarily associated not with the compensation of movements, which is the result of maladaptation, but with the restoration of motor function due to the activation of neuroplasticity processes in the motor and premotor cortex of the brain [51].

4. The current difficulties of using multisensor VR

Even though VR provides many unique advantages over traditional or new rehabilitation approaches, there are limitations to its widespread practical use as a routine rehabilitation method.

First, there is currently no sufficient evidence base for clinical studies that would demonstrate the unambiguous effectiveness of using VR in sensorimotor rehabilitation in various clinical groups in comparison with various traditional methods of motor rehabilitation. In addition, there is still quite a bit of information regarding the possibility of replacing physical exercises only with classes in virtual reality, namely, how interchangeable, and acceptable it is for short-term and long-term results of motor rehabilitation. That is why it is still impossible to say unambiguously how high the advantages of sensorimotor rehabilitation in VR are relative to those in the real physical world. Thus, all these questions justify the need to continue research in the field of studying the possibility of expanding the use of VR, as well as studying the short-term and long-term effectiveness of using VR in sensorimotor or cognitive rehabilitation, by accumulating a clinical base and obtaining the possibility of conducting a meta-analysis of research data to achieve the maximum high level of evidence. And although nowadays there are several studies in which attempts have been made to solve these problems, rehabilitation using VR technology continues to be considered only as an adjuvant method of sensorimotor and cognitive rehabilitation [5, 52].

The second important reason for the difficulty in the routine use of rehabilitation in VR is the relative high cost of equipment for using these systems as a method of rehabilitation within the framework of the telemedicine concept. A few years ago, the equipment needed to simulate VR, making it difficult to use for more than 40 minutes due to the heavy weight of the VR helmet, has become much more convenient today, because there has been an abrupt growth in the number of manufacturers of these technical devices offering more and more comfortable products for use.

5. Neuropasticity and VR

According to data obtained on biological models that allow studying the processes of neuroplasticity, the lack of stimulation of the motor cortex during the "critical period", which usually corresponds to an acute state after damage of any genesis, leads to the loss of corticospinal synaptic connections [53], while stimulation motor cortical networks in the same "critical period" may contribute to the partial restoration of some of these lost connections [54]. Long-term lack of stimulation of the motor cortex in the acute period after injury ultimately leads to the consolidation of existing changes that will prevent further restoration of function.

The key component of the theory of neuroplasticity activation is the dynamic nature of changes in neural connections during motor rehabilitation using VR technology, which can be adapted to the individual needs of the person being rehabilitated, providing a personalized approach to sensorimotor and cognitive rehabilitation.

Some studies demonstrate rich intrahemispheric cortical–cortical connections that link the occipital, parietal and frontal cortex [32, 33, 55]. Moreover, individual studies demonstrate that a significant number of motor neurons in the premotor and motor cortical areas are modulated by visual information [35, 37–39], suggesting that visual information can be a powerful signal for functional reorganization of sensorimotor connections.

The main parameters through which the processes of neuroplasticity can be activated, and which can be influenced when the immersive environment is formed, are visualized movement, biofeedback, motivation and learning through observation.

Rehabilitation measures in VR can also contribute to the process of functional reorganization in the CNS due to the activation of neuroplasticity processes.

The possibility of obtaining a significant rehabilitation result is achievable only with a long training process, because the formation of new skills, which is due to the activation of the processes of synaptogenesis or Hebb learning, as well as other mechanisms of neuroplasticity, does not give an immediate stable result, since stabilization requires subsequent reinforcement in order to stimulate the transition of interneuronal interaction from functional to morphologically fixed changes [56, 57].

Thus, in an immersive environment, it is quite easy to set the proper volume of tasks to be performed and combine them with secondary cognitive tasks, making the performance of a motor task interesting due to diversity, increasing the motivation of the person being rehabilitated for a long rehabilitation process [58].

Studies on biological models have demonstrated that the intensity and duration of physical exercise is one of the determining factors that have a significant impact on neuroplastic changes during rehabilitation [59]. For example, changes at the synaptic level in a biological experiment occurred after the animal was exposed to thousands of repetitions of a given task over a short period of time, i.e. 12,000 repetitions over 2–3 days [60, 61].

Also, it was noted that patients with CNS disease receiving rehabilitation on this occasion require more and more intensity of physical exercises in order to achieve positive results in restoring physical function, in comparison with the process of developing new motor skills in healthy people [62]. The duration of rehabilitation sessions to achieve positive effects in the restoration of function, for example, the upper limb, after a stroke also depends on the stage of stroke: from 1–2 hours in the acute stage [63] and up to 10–20 repetitions per training in the chronic stage of stroke [64]. At the same time, it is implied that during the entire time of the training, it is required to maintain a high level of motivation to achieve a positive result and maintain it throughout the entire course of rehabilitation.

The flexibility of most VR applications suggests that learning in a meaningful, enriched environment can be started earlier in recovery from an emerging CNS disease, such as a traumatic brain injury, compared to conventional exercise. An early start increases the rehabilitation potential by influencing neuroplastic processes and ensuring the activation of latent connections and cortical structures, which is also necessary to prevent the onset and progression of functional maladaptive processes. The same statement is relevant for patients with acute cerebrovascular accident, where verticalization in the first days after a stroke is limited due to pronounced concomitant pathology, which is usually the cause of the stroke, or the severity of the patient's movement disorders.

The possibility of automating the rehabilitation process in VR makes training more accessible for patients, and in the future can be used in telemedicine [65].

The hypothesis underlying the substantiation of the effectiveness of motor rehabilitation says that the success of motor learning occurs only at the moment of the maximum approximation of the rehabilitation exercise to real motor skills, which the rehabilitated person will use in the future in the real physical world, as well as when using neuro-feedback [57, 66].

In addition to using a simulated VR environment to restore basic movements or simple functions necessary to perform everyday household tasks, VR can become a training platform for developing patients' skills in using various means of individual rehabilitation, for example, for teaching the use of a motorized wheelchair or driving a car, etc. [67].

6. Implementation of neurofeedback in VR

Optimization of training and an increase in its effectiveness arises with the meaningfulness of the exercises performed and their optimal complication [58]. As a rule, only one repetition for training is not enough, since the formation of a motor skill occurs because of multiple repetitions with their use in solving real physical motion problems [68, 69]. For this, it is necessary to achieve an optimal combination of cognitive efforts required by the patient to solve motor problems during repetition of movements, and the complexity of the rehabilitation task.

The use of neuro-feedback contributes to an increase in the activation of structures that are usually not involved in the implementation of the performed movement in the norm. For example, in experiments on healthy subjects, it was demonstrated that the addition of neuro-feedback when performing a movement in VR leads to a more significant activation of the contrateral sensorimotor cortex according to the motor evoked potential [40]. In studies conducted in patients with hemiplegic infantile cerebral palsy, VR rehabilitation has demonstrated bilateral activation of the sensorimotor cortex and ipsilateral activation of the premotor cortex. After the completion of rehabilitation, bilateral activation disappeared, and the contralateral sensorimotor cortex continued to maintain a high level of activity [70].

Increasing the efficiency of neuro-feedback through the use of sensory channels is the most promising way to increase the possibility of motor learning using VR and ensure sufficient cognitive immersion in the VR environment.

The effectiveness of neuro-feedback can be assessed by such a parameter as productivity, which characterizes the quality of the movement performed by a person. The neural feedback obtained directly in the process of performing a rehabilitation exercise to restore motor function, after the completed rehabilitation task, can act as a criterion for evaluating the effectiveness. Some studies demonstrate that the enhancement of neurofeedback has an additional value in increasing the effectiveness of motor rehabilitation in patients after acute cerebrovascular accident [71]. At the same time, the direct implementation of neuro-feedback in the process of performing the rehabilitation task will be more promising, because on the basis of this approach, it is possible not only to visualize the performed movement and its quality, but also to carry out additional motivation of the patient by, for example, completing the construction of the full range of motion with a pronounced motor deficit in paralyzed limbs, or visualizing such a movement, which is the patient's in principle unable to perform (such as walking or running).

Perhaps, in some cases, this will cause a certain dissonance between real proprioceptive sensations and visual information provided to the rehabilitated person. However, in the end it will be perceived by the subject only from the positive side, since will allow him to demonstrate his independence and the ability to perform all the same actions without taking into account the existing disabling state. Such additional motivation will lead to the fact that the person being rehabilitated will be more motivationally involved in the process of restoring motor function, which will lead to better results in restoring motor function both in the short and long term.

As a complement to neuro-feedback, mainly implemented through the visual analyzer, VR provides the ability to use auditory and proprioceptive feedback, which are intuitively interpreted and implemented in real time, but with increased accuracy and consistency compared to the stimuli available in the physical world [4, 72].

The use of this technology as a supplement to visual information in an immersive environment through the activation of additional sensory systems also makes it possible to increase the degree of cognitive and emotional immersion in the VR environment and the task performed in it. This is especially in demand in patients with a certain damage to one or another sensory system at a different level from the peripheral part of the sensory analyzer to the cortical representation. It does not matter whether this damage arose because of a real disease, was acquired by the patient earlier, or was congenital. Thus, it is possible to achieve a more complete sensory saturation and get the maximum effect on the motor and premotor regions of the frontal and parietal lobes of the cerebral cortex.

These effects make it possible to neutralize sensory deprivation, which is observed in a patient after a pathological condition has arisen with gross damage to the CNS and manifests itself in pronounced motor disorders. Such patients are usually bedridden or wheelchair-bound and do not receive in full all those sensory sensations that a person experiences while freely moving in the physical world without physical limitations. Long-term sensory deprivation ultimately leads to neurotransmitter rearrangements, the clinical manifestation of which may be not only difficulty in restoring stato-locomotor function, but also the development of cognitive and emotional-volitional disorders.

Thus, rehabilitation measures, which are based on the activation of neuroplastic processes in the CNS after its damage, can be sufficiently fully modeled in an immersive environment, and multisensory neuro-feedback allows us to model the process of interaction with the VR environment as realistic and efficient as possible, which will contribute to solving most tasks.

7. Additional prospects for using VR

Also, it should be noted that the concept of motor learning forms the basis for the scientific substantiation of the integration of vocational training into rehabilitation practice, which will expand the possibilities of social adaptation of patients with a disabling disease, will contribute to their subsequent professional integration through training in professional activities, taking into account the existing motor or sensory deficit. This concept makes it possible to use VR to model specific conditions for teaching patients with movement disorders for their subsequent professional integration into society.

Today this problem is quite urgent, since any professional training is primarily focused on patients with intact motor impairments or their minimal severity.

In the context of the wide possibilities of VR for modeling a wide variety of conditions and tasks using various sensory channels, it can be assumed that its use will be in demand for the rapid formation of the necessary environment, which allows to restore not only any lost skill due to the developed disease, but also for the training of professional skills in patients with pre-existing movement or other disabilities.

An important factor influencing the improvement of the effectiveness of rehabilitation in VR is the possibility of its use as remote rehabilitation within the framework of the telemedicine concept.

Such a combination is possible only in conditions where a clear assessment of biometrics is available, which of course is possible in VR. The use of artificial intelligence algorithms will allow automating the process of motor rehabilitation, taking into account the initial personified clinical data.

Effective approaches to rehabilitation should include targeted training, individual feedback based on multisensory interaction, an individual exercise schedule, a fairly long and frequent repetition of motor exercises, exciting game scenarios, individual rehabilitation programs taking into account the characteristics of motor deficits and individual preferences of the person being rehabilitated [73].

Attention should be paid to the need to further form the evidence base for assessing the effects of rehabilitation in VR and the contribution of individual elements that form neuro-feedback through various sensory channels and primarily through the proprioceptive channel, which allows the formation of explicit interaction with VR objects, identical to the interaction with physical objects [74–76].

8. Conclusion

The spread of the use of VR in rehabilitation practice is determined by numerous factors, including the technical availability of equipment and software, the possibility of creating personalized rehabilitation exercises to achieve a higher rehabilitation effect.

The use of a multisensory component in the implementation of neuro-feedback allows one to achieve potentially better results in motor rehabilitation. Proprioception, as one of the components of such neuro-feedback, is the most promising way of forming sensations that are as close as possible to natural sensations obtained during physical contact with objects of the real world.

It is possible to single out the factors that determine the more effective use of VR for solving various rehabilitation tasks: immersivity, neuro-feedback, the possibility of multiple repetition of a motor task with visualization of such a movement, the use of artificial intelligence and mathematical models on the basis of which the movement is visualized, taking into account all the richness of kinematics and synergy of such movements, as well as the possibility of objective continuous monitoring of the entire rehabilitation process.

The use of devices that create proprioceptive sensations upon contact with VR objects make it possible to count on obtaining explicit VR environments, which will have a much wider range of users. This gives reason to count on achieving maximum efficiency in the restoration of motor functions during the rehabilitation and habilitation of patients with various functional disorders, both as a result of acquired pathology and as a result of disorders that have arisen at various stages of ontogenesis.

It is extremely important today to form a complete understanding of how different sensory and tactile manipulations in VR affect the dynamics of various processes in the CNS, since the study of this issue will reveal the full potential of rehabilitation through VR.

One of the important, unsolved problem, possibly allowing to reveal the full potential of rehabilitation, is the formation of a complete understanding of how different sensory influences in VR affect the dynamics of various processes in the central nervous system, including the dynamics of sensorimotor connections. Understanding of this effect will make it possible to achieve greater personalization of the rehabilitation process, not only based on the severity of a motor defect, but also considering functional disorders in a complex, multicomponent, hierarchically arranged motor system.

Acknowledgements

SamSMU students, and staff Samara regional hospital after D.V. Seredavin.

Conflict of interest

The results were obtained in Leading Research Centre supported by RF Ministry of Communications and RVC JSC fund (Grant No. 003/20 03-17-2020, Grant ID - 0000000007119P190002).

Notes/thanks/other declarations

We thank management Samara Medical University for the opportunity to conduct scientific work.

Author details

Alexander Vladimirovich Zakharov[1]*, Alexander Vladimirovich Kolsanov[2], Elena Viktorovna Khivintseva[1], Vasiliy Fedorovich Pyatin[3] and Alexander Vladimirovich Yashkov[4]

1 Department of Neurology and Neurosurgery, Samara Medical University, Samara, Russia

2 Department Operative Surgery and Clinical Anatomy with Innovative Technology Course, Samara Medical University, Samara, Russia

3 Department of Physiology with a Course on Life Safety and Disaster Medicine, Samara Medical University, Samara, Russia

4 Department of Medical Rehabilitation, Sports Medicine, Physiotherapy and Balneology, Samara Medical University, Samara, Russia

*Address all correspondence to: zakharov1977@mail.ru

References

[1] Stillman BC. Making Sense of Proprioception: The Meaning of Proprioception, Kinaesthesia and Related Terms. Physiotherapy. 2002;88(11):667-676. DOI: 10.1016/S0031-9406(05)60109-5

[2] Pereira EM, Rueda FM, Diego IA, De La Cuerda RC, De Mauro A, Page JM. Use of Virtual Reality Systems as Proprioception Method in Cerebral Palsy: Clinical Practice Guideline. Neurologı'a (English Edition). 2014;29(9):550-559. DOI:10.1016/j.nrleng.2011.12.011.

[3] Weiss PL, Rand D, Katz N, Kizony R. Video capture virtual reality as a flexible and effective rehabilitation tool. J Neuroeng Rehabil. 2004;1:12. DOI: 10.1186/1743-0003-1-12.

[4] Holden MK. Virtual environments for motor rehabilitation: review. Cyberpsychol Behav 2005;8:187-211. discussion 212-189. DOI: 10.1089/cpb.2005.8.187.

[5] Laver KE, Lange B, George S, Deutsch JE, Saposnik G, Crotty M. Virtual reality for stroke rehabilitation. Cochrane Database of Systematic Reviews. 2017;11(11):CD008349. DOI: 10.1002/14651858.CD008349.pub4

[6] Maier M, Ballester BR, Duff A, Oller ED, Verschure PFMJ. Effect of Specific Over Nonspecific VR-Based Rehabilitation on Poststroke Motor Recovery: A Systematic Meta-analysis. Neurorehabilitation and Neural Repair. 2019;33:112-129. DOI: 10.1177/1545968318820169.

[7] Fulk GD. Locomotor training and virtual reality-based balance training for an individual with multiple sclerosis: a case report. J Neurol Phys Ther. 2005;29(1):34-42. DOI: 10.1097 / 01.npt.0000282260.59078.e4.

[8] Rose FD, Brooks BM, Rizzo AA. Virtual reality in brain damage rehabilitation: Review. CyberPsychology & Behaviour. 2005;8(3):241-271. DOI: 10.1089 / cpb.2005.8.241.

[9] Piron L, Tombolini P, Turolla P, Zucconi C, Agostini M, Dam M, et al. Reinforced feedback in virtual environment facilitates the arm motor recovery in patients after a recent stroke. Virtual Rehabilitation. 2007; 2007:121-123. DOI: 10.1109/ICVR.2007.4362151.

[10] Merians AS, Fluet GG, Qiu Q, Lafond I, Adamovich SV. Learning in a virtual environment using haptic systems for movement re-education: Can this medium be used from remodeling other behaviors and actions? Journal of Diabetes Science and Technology. 2011;5(2):301-308. DOI:10.1177/193229681100500215.

[11] Bohil CJ, Alicea B, Biocca FA. Virtual reality in neuroscience research and therapy. Nature Reviews. Neuroscience. 2011;12(12):752-762. DOI: 10.1038 / nrn3122.

[12] Adams H, Narasimham G, Rieser J, Creem-Regehr S, Stefanucci J, Bodenheimer B. Recalibration and Prism Adaptation of Children and Teens in Immersive Virtual Environments. IEEE Transactions on Visualization & Computer Graphics. 2018;24(4):1408-1417. DOI: 10.1109/ TVCG.2018.2794072.

[13] Bodenheimer B, Creem-Regehr S, Stefanucci J, Shemetova E, Thompson WB. Prism Aftereffects for Throwing with a Self-Avatar in an Immersive Virtual Environment. In: Rosenberg ES. (eds.) 2017 IEEE Virtual Reality (VR) Proceedings: March 18-22, 2017, Los Angeles, CA, USA. Piscataway, NJ: IEEE; 2017. p. 141-147.

[14] Merians AS, Poizner H, Boian R, Burdea G, Adamovich S. Sensorimotor

training in a virtual reality environment: Does it improve functional recovery poststroke. Neurorehabilitation and Neural Repair. 2006;20(2):252-267. DOI: 10.1177/1545968306286914.

[15] Zakharov AV, Bulanov VA, Khivintseva EV, Kolsanov AV, Bushkova YV, Ivanova GE. Stroke Affected Lower Limbs Rehabilitation Combining Virtual Reality With Tactile Feedback. Frontiers in robotics and AI. 2020; 7:81. DOI: 10.3389/frobt.2020.00081.

[16] You SH, Jang SH, Kim YH, Hallet M, Ahn SH, Kwon YH, et al. Virtual reality-induced cortical reorganization and associated locomotor recovery in chronic stroke: an experimenter-blind randomized study. Stroke. 2005;36(6):1166-1171. DOI: 10.1161 / 01.STR.0000162715.43417.91.

[17] Boyd LA, Randhawa B, Vidoni ED, Wessel BD. Motor learning after stroke: Is skill acquisition a prerequisite for contralesional neuroplastic change? Neuroscience Letters. 2010;482(1):21-25. DOI: 10.1016 / j.neulet.2010.06.082.

[18] Bray S, Shimojo S, O'Doherty JP. Direct instrumental conditioning of neural activity using functional magnetic resonance imaging-derived reward feedback. Journal of Neuroscience. 2007;27(28):7498-7507. DOI: 10.1523/JNEUROSCI.2118-07.2007.

[19] Hadipour-Niktarash A, Lee CK, Desmond JE, Shadmehret R. Impairment of retention but not acquisition of a visuomotor skill through time-dependent disruption of primary motor cortex. Journal of Neuroscience. 2007;27(49):13413-13419. DOI:10.1523/JNEUROSCI.2570-07.2007.

[20] Muellbacher W, Ziemann U, Boroojerdi B, Cohen L, Hallett M. Role of the human cortex in rapid motor learning. Experimental Brain Research. 2001;136(4):431-438. DOI: 10.1007/s002210000614.

[21] Muellbacher W, Ziemann U, Wissel J, Dang N, Kofl er M, Facchini S, et al. Early consolidation in human primary motor cortex. Nature. 2002;415(6872):640-644. DOI: 10.1038/nature712.

[22] Richardson AG, Overduin SA, Valero-Cabré A, Padoa-Schioppa C, Pascual-Leone A, Bizzi E, et al. Disruption of primary motor cortex before learning impairs memory of movement dynamics. Journal of Neuroscience. 2006;26(48):12466-12470. DOI: 10.1523/JNEUROSCI.1139-06.2006.

[23] Wise SP, Moody SL., Blomstrom KJ, Mitz AR. Changes in motor cortical activity during visuomotor adaptation. Experimental Brain Research. 1998;121(3):285-299. DOI: 10.1007 / s002210050462.

[24] Leonard G, Tremblay F. Corticomotor facilitation associated with observation, imagery and imitation of hand actions: A comparative study in young and old adults. Experimental Brain Research. 2007;177(2):167-175. DOI: 10.1007 / s00221-006-0657-6.

[25] Patuzzo S, Fiaschi A, Manganotti P. Modulation of motor cortex excitability in the left hemisphere during action observation: A single- and paired-pulse transcranial magnetic stimulation study of self- and non-self-action observation. Neuropsychologia. 2003;41 (9):1272-1278. DOI: 10.1016/s0028-3932(02)00293-2.

[26] Stefan K, Cohen LG, Duque J, Mazzocchio R, Celnik P, Sawaki L, et al. Formation of a motor memory by action observation. Journal of Neuroscience. 2005;25(41):9339-9346. DOI: 10.1523/JNEUROSCI.2282-05.2005.

[27] Strafella AP, Paus T. Modulation of cortical excitability during action observation: A transcranial magnetic stimulation study. Neuroreport.

2000;11(10):2289-2292. DOI: 10.1097 / 00001756-200007140-00044.

[28] Bulanov VA, Zakharov AV, Chaplygin SS. Solving classification problems of visual evoked potentials for the brain-computer interfaces. 2020 IOP Conf. Ser.: Mater. Sci. Eng. 2020; 862: 052051. DOI:10.1088/1757-899X/862/5/052051.

[29] Bulanov VA, Zakharov AV, Khivintseva EV. Wavelet transform for the identification of P300. 2020 IOP Conf. Ser.: Mater. Sci. Eng. 2020;862:052049. DOI:10.1088/1757-899X/862/5/052049.

[30] Dum RP, Strick PL. Frontal lobe inputs to the digit representations of the motor areas on the lateral surface of the hemisphere. Journal of Neuroscience. 2005;25(6):1375-1386. DOI: 10.1523 / JNEUROSCI.3902-04.2005.

[31] Fang PC, Stepniewska I, Kass JH. Ipsilateral cortical connections of motor, premotor, frontal eye, and posterior parietal fields in a prosimian primate, Otolemur garnetti. The Journal of Comparative Neuroscience. 2005;490(3):305-333. DOI:10.1002/ cne.20665.

[32] Lewis SJG, Slabosz A, Robbins TW, Barker RA, Owen AM. Dopaminergic basis for deficits in working memory but not attentioanl set-shifting in Parkinson's disease. Neuropsychologia. 2005;43(6):823-832. DOI: 10.1016 / j.neuropsychologia.2004.10.001

[33] Lewis JW, Van Essen DC. Corticocortical connections of visual, sensorimotor, and multimodal processing areas in the parietal lobe of the macaque monkey. The Journal of Comparative Neurology. 2000;428(1):112-137. DOI: 10.1002 / 1096-9861 (20001204) 428: 1 <112 :: aid-cne8> 3.0.co; 2-9.

[34] Lewis JW, Van Essen DC. Mapping of architectonic subdivisions in the macaque monkey, with emphasis on parieto-occipital cortex. The Journal of Comparative Neurology. 2000;428(1):79-111. DOI: 10.1002/ 1096-9861(20001204)428:1<79::aid-cne7>3.0.co;2-q.

[35] Graziano MS. Where is my arm? The relative role of vision and proprioception in the neuronal representation of limb position. Proceedings of the National Academy of Science. 1999;96(18): 10418-10421. DOI: 10.1073 / pnas.96.18.10418.

[36] Graziano MS, Gandhi S. Location of the polysensory zone in the precentral gyrus of anesthetized monkeys. Experimental Brain Research. 2000;135(2): 259-266. DOI: 10.1007 / s002210000518.

[37] Graziano MS, Gross CG. Spatial maps for the control of movement. Current Opinion in Neurobiology. 1998; 8(2):195-201. DOI: 10.1016 / s0959-4388 (98) 80140-2.

[38] Graziano MS, Gross CG. Visual responses with and without fixation: Neurons in premotor cortex encode spatial locations independently of eye position. Experimental Brain Research. 1998;118 (3):373-380. DOI: 10.1007/ s002210050291.

[39] Kakei S, Hoffman DS, Strick PL. Sensorimotor transformations in cortical motor areas. Neuroscience Research. 2003; 46(1): 1-10. DOI: 10.1016 / s0168-0102 (03) 00031-2.

[40] Adamovich SV, August K, Merians A, Tunik E. A virtual reality-based system integrated with fmri to study neural mechanisms of action observation-execution: A proof of concept study. Restorative Neurology and Neuroscience. 2009;27(3): 209-223. DOI: 10.3233 / RNN-2009-0471.

[41] Snijders HJ, Holmes NP, Spence C. Direction-dependent integration

of vision and proprioception in reaching under the infl uence of the mirror illusion. Neuropsychologia. 2007;45(3):496-505. DOI: 10.1016 / j.neuropsychologia.2006.01.003.

[42] Riva G. Virtual environments in neuroscience. IEEE Transactions of Information Technology of Biomedicine. 1998;2(4):275-281. DOI: 10.1109/4233.737583.

[43] Riva G, Castelnuovo G, Mantovani F. Transformation of flow in rehabilitation: The role of advanced communication technologies. Behavioural Research Methods. 2006;38(2):237-244. DOI: 10.3758 / BF03192775.

[44] Zakharov AV, Khivintseva EV, Pytin VF, Kolsanov AV, Kalinin VA, Osadchuk MA, et al. Restoration of motive function of the lower extremities using virtual reality technique. Journal of Advanced Pharmacy Education & Research. 2019;9(2):102-107.

[45] Deutsch JE, Merians AS, Adamovich S, Poizner H, Budrea GC. Development and application of virtual reality technology to improve hand use and gait of individuals poststroke. Restorative Neurology Neuroscience. 2004;22(3-5):371-386.

[46] Kenyon RV, Afenya MB. Training in virtual and real environments. Annals of Biomedical Engineering. 1995;23(4):445-455. DOI: 10.1007 / BF02584444.

[47] Merians AS, Jack D, Boian R, Tremaine M, Burdea GC, Adamovich SV, et al. Virtual reality-augmented rehabilitation for patients following stroke. Physical Therapy. 2002;82(9): 898-915. DOI:10.1093/ptj/82.9.898.

[48] Carey LM, Abbott DF, Puce A, Jackson GD, Syngeniotis A, Donnan GA. Reemergence of activation with poststroke somatosensory recovery: A serial fMRI case study. Neurology. 2002;59(5):749-752. DOI: 10.1212/wnl.59.5.749.

[49] Carey LM, Abbott DF, Egan GF, O'Keefe GJ, Jackson GD, Bernhardt J, et al. Evolution of brain activation with good and poor motor recovery after stroke. Neurorehabilitation and Neural Repair.2006;20(1):24-41. DOI: 10.1177 / 1545968305283053.

[50] Small SL, Hlustik P, Noll DC, Genovese C, Solodkin A. Cerellar hemispheric activation ipsilateral to the paretic hand correlated with functional recovery after stroke. Brain. 2002;125(7):1544-1557. DOI: 10.1093 / brain / awf148.

[51] Subramanian SK, Lourenço CB, Chilingaryan G, Sveistrup H, Levin MF. Arm motor recovery using a virtual reality intervention in chronic stroke: Randomized control trial. Neurorehabilitation and Neural Repair. 2013;27(1):13-23. DOI: 10.1177 / 1545968312449695.

[52] Gibbons EM, Thomson AN, Noronha MD, Joseph S. Are virtual reality technologies effective in improving lower limb outcomes for patients following stroke–a systematic review with metaanalysis. Top. Stroke Rehabil. 2016;23(6):440-457. DOI: 10.1080/10749357.2016.11 83349.

[53] Friel KM, Barbay S, Frost SB, Plautz EJ, Stowe AM, Dancause N, et al. Effects of a rostral motor cortex lesion on primary motor cortex hand representation topography in primates. Neurorehabil Neural Repair. 2007;21:51-61. DOI: 10.1177/1545968306291851.

[54] Salimi I, Friel KM, Martin JH. Pyramidal tract stimulation restores normal corticospinal tract connections and visuomotor skill after early postnatal motor cortex activity blockade. J Neurosci.

2008;28(29):7426-7434. DOI: 10.1523 / JNEUROSCI.1078-08.2008.

[55] Stepniewska I, Fang PC, Kaas JH. Microstimulation reveals specialized subregions for different complex movements in posterior parietal cortex of prosimian galagos. Proc Natl Acad Sci U S A. 2005;102(13):4878-4883. DOI: 10.1073/pnas.0501048102.

[56] Adamovich SV, Fluet GG, Tunik E, Merians AS. Sensorimotor training in virtual reality: A review. Neurorehabilitation. 2009;25(1):29-44. DOI: 10.3233 / NRE-2009-0497.

[57] Schmidt RA, Lee TD. Motor control and learning: A behavioral emphasis (5th ed.). Champaign, IL: Human Kinetics; 2011.

[58] Kleim J, Jones T. Principles of experience-dependent neural plasticity: Implications for rehabilitation after brain damage. Journal of Speech, Language & Hearing Research. 2008;51(1):S225–S239. DOI: 10.1044/1092-4388(2008/018).

[59] Krakauer JW, Carmichael ST, Corbett D, Wittenberg GF. Getting neurorehabilitation right: What can be learned from animal models? Neurorehabilitation and Neural Repair. 2012;26(8):923-931. DOI: 10.1177/1545968312440745.

[60] Nudo RJ, Milliken GW. Reorganization of movement representations in primary motor cortex following focal ischemic infarcts in adult squirrel monkeys. Journal Neurophysiol. 1996;75(5):2144-2149. DOI: 10.1152 / jn.1996.75.5.2144.

[61] Plautz EJ, Milliken GW, Nudo RJ. Effects of repetitive motor training on movement representations in adult squirrel monkeys: role of use versus learning. Neurobiol Learn Mem. 2000;74(1):27-55. DOI: 10.1006/ nlme.1999.3934.

[62] Cirstea MC, Ptito A, Levin MF. Arm reaching improvements with short-term practice depend on the severity of the motor deficit in stroke. Experimental Brain Research. 2003;152(4):476-488. DOI: 10.1007/s00221-003-1568-4.

[63] Bernhardt J, Dewey H, Thrift A, Donnan G. Inactive and alone: physical activity within the first 14 days of acute stroke unit care. Stroke. 2004;35(4):1005-1009. DOI: 10.1161/01. STR.0000120727.40792.40.

[64] Lang C, Macdonald J, Gnip C. Counting repetitions: an observational study of outpatient therapy for people with hemiparesis post-stroke. J Neurol Phys Ther. 2007;31(1):3-11. DOI: 10.1097/01.npt.0000260568.31746.34.

[65] Holden M, Todorov E, Callahan J, Bizzi E. Virtual environment training improves motor performance in two patients with stroke: Case report. Neurology Report. 1999;23(2):57-67. DOI: 10.1097/01253086-199923020-00013.

[66] Barnett ML, Ross D, Schmidt RA, Todd B. Motor skills learning and the specificity of training principle. Research Quarterly. 1973;44(4):440-447. DOI: 10.1080/10671188.1973.10615224.

[67] Kizony R, Levin MF, Hughey L, Perez C, Fung J. Cognitive load and dual-task performance during locomotion poststroke: A feasibility study using a functional virtual environment. Physical Therapy. 2010;90(2):252-260. DOI: 10.2522/ ptj.20090061.

[68] Lee TD, Swinnen SP, Serrien DJ. Cognitive effort and motor learning. QUEST. 1994;46: 328-344. DOI: 10.1080/00336297.1994.10484130.

[69] Lehto NK, Marley TL, Ezekiel HJ, Wishart LR, Lee TD, Jarus T. Application of motor learning

principles: The physiotherapy client as a problem-solver. IV. Future directions. Physiotherapy Canad. 2001;53(2):109-114.

[70] You SH, Jang SH, Kim YH, Kwon YH, Barrow I, Hallett M. Cortical reorganization induced by virtual reality therapy in a child with hemiparetic cerebral palsy. Developmental Medicine & Child Neurology. 2005;47:628-635. DOI: 10.1111/j.1469-8749.2005.tb01216.x.

[71] Molier BI, Van Asseldonk EH, Hermens HJ, Jannink MJ. Nature, timing, frequency and type of augmented feedback; does it influence motor relearning of the hemiparetic arm after stroke? A systematic review. Disability and Rehabilitation. 2010;32(22):1799-1809. DOI: 10.3109/09638281003734359.

[72] Subramanian SK, Massie CL, Malcolm MP, Levin MF. Does provision of extrinsic feedback result in improved motor learning in the upper limb poststroke? A systematic review of the evidence. Neurorehabilitation Neural Repair. 2010;24(2):113-124. DOI: 10.1177/1545968309349941.

[73] Timmermans AA, Seelen HA, Willmann RD, Kingma H. Technology-assisted training of arm-hand skills in stroke: Concepts on reacquisition of motor control and therapist guidelines for rehabilitation technology design. Journal of Neuroengineering & Rehabilitation. 2009;6:1. DOI: 10.1186/1743-0003-6-1.

[74] Rose FD, Attree EA, Brooks BM, Johnson DA. Virtual environments in brain damage rehabilitation: A rationale from basic neuroscience. In Riva G, Wiederhold BK, Molinari M. (eds.). Virtual environments in clinical psychology and neuroscience. Amsterdam, The Netherlands: Ios Press; 1998. p. 233-242.

[75] Rose FD, Attree EA, Brooks BM, Parslow DM, Penn PR, Ambihaipahan N. Training in virtual environments: Transfer to real world tasks and equivalence to real task training. Ergonomics. 2000;43(4):494-511. DOI: 10.1080/001401300184378.

[76] Sandlund M, McDonough S, Hager-Ross C. Interactive computer play in rehabilitation of children with sensorimotor disorders: A systematic review. Developmental Medicine & Child Neurology. 2009;51(3):173-179. DOI: 10.1111 / j.1469-8749.2008.03184.x.